DEFENDING EVOLUTION
IN THE CLASSROOM

A guide to the creation/evolution controversy

Brian J. Alters
Sandra M. Alters

JONES AND BARTLETT PUBLISHERS
Sudbury, Massachusetts
BOSTON   TORONTO   LONDON   SINGAPORE

World Headquarters
Jones and Bartlett Publishers
40 Tall Pine Drive
Sudbury, MA 01776
978-443-5000
info@jbpub.com
www.jbpub.com

Jones and Bartlett Publishers Canada
2406 Nikanna Road
Mississauga, ON L5C 2W6
CANADA

Jones and Bartlett Publishers International
Barb House, Barb Mews
London W6 7PA
UK

Library of Congress Cataloging-in-Publication Data

Alters, Brian J.
    Defending evolution in the classroom: a guide to the creation/evolution controversy/
Brian J. Alters, Sandra M. Alters.
        p. cm.
        ISBN 0-7637-1923-4 (hardcover)—ISBN 0-7637-1118-7 (pbk.)
        1. Evolution (Biology)--Study and teaching. 2. Evolution (Biology)--Philosophy. 3.
Creationism. I. Alters, Sandra. II. Title.

QH362 .A62 2001
576.8'071--dc21

2001029193

Credits
V.P., Senior Managing Editor: Judith H. Hauck
Associate Editor: Victoria Jones
Special Projects Editor: Mary Hill
Copyeditor: Kathy Smith
Composition and Design: Thompson Steele, Inc.
Cover Design: Anne Spencer
Book Manufacture: Malloy Lithographers
Cover Manufacture: John Pow Company

On the cover: *Homo erectus* skull, Peking Man, 400,000–500,000 YA. Also known as *Pithecanthropus pekinensis (Sinanthropus).* Image of this recreation courtesy of Bone Clones® (www.boneclones.com).

Printed in the United States
04   03   02   01   10 9 8 7 6 5 4 3 2 1

*Dedicated to instructors
who teach evolution
despite the pressures to do otherwise.*

About the Authors

Brian J. Alters, Ph.D., lead author, holds appointments internationally at McGill University, Montreal, where he was recently named William Dawson Scholar, and at Harvard University, Cambridge. Dr. Alters is Director of the newly created Evolution Education Research Centre.

Sandra M. Alters, Ph.D., is a former Associate Professor of Biology and Education at the University of Missouri-St. Louis. Dr. Alters has authored four science textbooks in addition to her most recent, *Understanding Life,* and has written over 60 chapters and features in books about biology, education, and science education.

Contents

Foreword

I have often, in my writings, deplored our all-too-human tendency to dichotomize complex issues into an overly simplified contrast of us against them—the good guys *vs.* the bad guys. Nonetheless, at least in the arena of proper response to social or political struggles, integrity often demands that we clearly advocate one side of a dispute, even while we strive to understand the complexity of motives, and the range of beliefs, among our adversaries. Scientists and educators must give a clear and uncompromising response to long-standing attempts by creationists either to eliminate (or seriously to adulterate) the teaching of evolution, or to insert, by legal or social fiat, the nonscientific "alternative" of oxymoronic "creation science" into the curriculum of public school science courses. We must resist these efforts with all our heart and force, for the very integrity of education hangs in the balance for two major reasons (that creationists have often managed to obfuscate in their public rhetoric):

1. Although science does not strive for unattainable certainty, the factuality of evolution is as well confirmed as any major discovery of science, including the heliocentric theory of the solar system, and the periodic table of chemical elements. Moreover, evolution is the core concept of all the biological sciences, not a peripheral concern that could be shelved for tactical reasons of educational peace. By the norms and ethics of their profession, scientists and educators cannot, for practical reasons of comfort or diplomacy, fail to teach the most important, and the most highly confirmed (not to mention wonderfully fascinating), concept of their chosen subject.

2. The content of so-called "creation science," as our professional scientific societies have documented *ad nauseam,* and as our courts have never failed to recognize in their final judgments, violates the procedures of scientific inquiry, and cannot be legitimately included within the subject matter of courses in science. (The motivations of those who have tried to insinuate creationism into curricula of public school science courses have been exposed as political, or narrowly religious—that is, as supportive of a particular and minority view of religion, and not of the opinions of most religious Americans, who do not oppose evolution—and not as originating from any scientific data.) Therefore, if creationism were taught in public school science courses, students would not only receive misinformation and poor education in biology, they would also be unable to understand the norms and procedures of the entire scientific enterprise.

Nonetheless, although we can state the nature of this required response with ease (a true "no brainer" for anyone dedicated to the integrity of science, and of intellectual life in general), the practical means for making this response in a maximally potent and effective manner can be elusive and complex—for creationism exists in many "flavors," and the battle must be fought on a person-to-person and classroom-by-classroom basis. This shift in the primary place of struggle emerges from our successes, not our failures. We have ultimately won every major legal battle on First Amendment grounds (as courts eventually recognize the partisan religious, rather than the scientific, motivations of creationists), and this constitutional imperative has also served us well in overt political battles (as in the ousting, in school board elections of 2000, of Kansas creationists who temporarily removed evolution from the list of required subjects for curricula of high school biology courses). Consequently, creationists have shifted their focus to personal pressure and persuasion applied to individual teachers in their particular classrooms. These "scare" tactics have, unfortunately, often succeeded, as teachers (feeling a lack of support, or simply disinclined to undergo stigmatization, or even danger) compromise their training and integrity by a range of accommodations—from not teaching evolution, to calling the subject by a different and euphemistic name, to teaching the material later in the course, or less intensively.

Because the battle has now shifted directly to the classroom as a primary target, we need, more than anything else at this point, to produce materials that individual teachers can use to understand, and effectively to combat (that is, with force, but also with respect and empathy for the sincere beliefs and sensibilities of creationist students) this serious and pervasive threat to excellence in education. Brian and Sandra Alters are distinguished professional educators with special expertise (and compassionate understanding for adversaries) in this contentious subject. (Brian heads an institute for the study of education in evolution at McGill University in Montreal; Sandra is the author of several distinguished textbooks in science.) This important and practical guide, written primarily for teachers in the trenches of this debate, discusses both the arguments and underpinnings of creationist beliefs, and also presents a series of suggestions and proven tactics for responding to the concerns of students. Thus, this book becomes a vital document in one of the most important issues of our age.

I particularly appreciate the authors' subtle understanding of the range of creationist beliefs and motivations—thus avoiding a harmful caricature of all creationists as Bible-thumping literalists committed to 6 days of creation (24 hours each), and to an earth no more than 10,000 years old. The complexity of the subject can best be captured in two statements, emphasized throughout by the Alters:

1. Most emphatically, creationism *vs.* evolution cannot be equated with religion *vs.* science. First of all, religion and science do not and cannot stand in genuine opposition, for each vital endeavor treats a different aspect of our complex existence (science as an enterprise devoted to discovering the factual character of nature; religion as a source of moral discussion, and a focus in our search to understand the meaning and purpose of our lives). Second, the vast majority of professional theologians and spiritual leaders recognize this separation, and accept evolution as a factual statement about the natural world, with no negative bearing upon religious inquiry. Religious Americans—a vast majority of our population—need not fear that evolution will undermine their theological concerns. Indeed, a high percentage of professional evolutionary biologists are devoutly religious in their personal lives.

2. As stated above, creationism spans an enormous range of beliefs and motivations—from Biblical literalists to "intelligent design" theorists who accept the ancient age of the earth and also allow subsequent evolution following the creation of life's basic complexity; and from political activists using creationism as a wedge to exploit a larger social goal, to open-minded students who have been raised in creationist families, but who are willing, even eager, to study alternatives. To be effective, teachers must vary their approaches and responses to meet this enormous variety of concerns, tactics, and motives among creationist students and community members.

This battle must be won, but we cannot prevail (or at least cannot prevail honorably) unless we meet our creationist questioners by grappling with their diversity of arguments, and with respect for the sincere and important reasons behind their misunderstanding of material that properly belongs within the domain of science, and cannot threaten the essence of religion.

Stephen Jay Gould
Museum of Comparative Zoology
Harvard University

Introduction

Deemphasize the teaching of evolution in science curricula?

Give creationism equal time
with evolution in the science classroom?

Teach that an intelligent designer
might be a better scientific explanation than evolution?

A research firm recently commissioned to study U.S. public opinion on matters relevant to the above questions stated in their report that, "Perhaps not since the Scopes Monkey Trial of the 1920s has so much public attention been paid to this [creation/evolution] issue."[1] Bolstering this statement are their results, showing that a significant majority of Americans (81%) say the issue of whether creationism should be taught in schools is important to them.[2] (The term "creationism" refers to explanations that account for the diversity of life on earth and that are contrary to evolution.[3])

When it comes to teaching and learning evolution—a fact of science—creationists claim that the instructors, books, and scientists are all inaccurate. Most science instructors would agree that this claim is blatantly wrong. Nonetheless, millions of people in the United States agree with it, and they have attended schools with noncreationist science curricula. In fact, approximately half of America appears to hold a view that is at odds with science. The results of various polls concerning people's acceptance or rejection of evolution generally report close to a 50% rejection rate.[4] These statistical data are not results of polling students alone; they are results of polling a cross-section of the populace, primarily those who have been educated in U.S. schools that use non-creationist science curricula.

What Americans think about their origins is often shocking to those of us who teach about evolution. Gallup polls report that almost 50% of Americans responded that "God created human beings pretty much in their present form at one time within the last 10,000 years or so"[5] and that almost 70% support teaching creationism in schools.[6] Results from a National Science Board countrywide quiz reveal that only about half of respondents answered "False" to the statement, "The earliest humans lived at the same time as the dinosaurs." A mere 45% responded "True" to "Human beings, as we know them today, developed from earlier species of animals."[7]

Regardless of whether people think we developed from earlier species of animals, many don't seem to know what the term *evolution* means. For example, another national poll reports that of those who recall ever having heard the term *evolution,* only 50% chose the correct layman's definition—and roughly one in three thinks "evolution means human beings have developed from apes."[8]

The poll also reports confusion with the term *theory* and confusion about or rejection of the factuality of evolution. With regard to theory, 74% agreed that "evolution is commonly referred to as the *theory* of evolution because it has not yet been proven scientifically."[9] On another question, only 29% felt that evolution was "completely" or "mostly accurate"; while the remaining 71% responded "mostly not accurate," "completely not accurate," "not sure," or "might or might not be accurate, you can never know for sure."[10] Interestingly, those who express the least certainty about evolution lack a college education.[11]

Misunderstanding evolution is not a new phenomenon; it has been the case despite decades of science curricula attempting to teach the subject. Moreover, this is not just an issue in the biological sciences; many types of creationism also contradict established concepts in other scientific disciplines such as astronomy, chemistry, geology, and physics. Biology instructors are not alone in this assault on science education.

Then, of course, there's the political side to the evolution/creation issue. In 1999, the Kansas State Board of Education voted to remove almost all mention of evolution from the state's education standards and assessments for public schools. Over 12 other states have fought similar versions of an anti-evolution battle, including some that have succumbed to placing disclaimers about evolution in their biology textbooks.[12] In the 2000 preliminary presidential campaigns, most of the candidates favored the position that *both* evolution and creationism be taught in schools and added that such decisions should be made at the local level. At that same time, leading anti-evolution organizations reported good news to their constituencies of having "consultations" with school boards and legislators, and the president of one of these organizations reported that "the creation message is getting out better than ever."[13]

In teaching evolution to our students, some science educators say we should just ignore the creationists and not address students' creationist-related concerns. This has been a popular method of teaching evolution at high schools and colleges for a long time, and continues to be. However, decades of ignoring creationism have not succeeded in significantly improving the public's understanding of evolution. Randy Moore, Editor of *The American Biology Teacher,* a leading journal of biological education, concluded:

> Creationists are more powerful than ever. They're winning, not in terms of court cases, but what happens in classrooms. I get three to five phone calls a week from teachers with problems.[14]

He asserts that a factor helping creationism to become so popular is the hesitancy of science instructors to resist it.[15]

While most Americans seem to misunderstand evolution and many reject its scientific factuality, an overwhelming majority (83%) want evolution taught in public schools, which is good news.[16] More good news is that ministers, priests, rabbis, and Pope John Paul II have acknowledged the validity of evolutionary science, while maintaining their religious beliefs.[17] (Of course, numerous evolutionary biologists are devoutly religious.) There are certainly nonreligious creationists (and they will be discussed), but they appear to be the vast minority and rarely fight against the teaching of evolution.

The creation/evolution controversy is not an academic argument between the disciplines of religion and science, but rather a clash between creationism and science, and, many would add, between religion and creationism. Personally, the two of us have a great respect for religion, with one raised as a Catholic and the other raised as a Protestant who has attended Christian schools and seminary. We pick no quarrel with anyone's theology or church traditions; we do not contest teaching creationism in homes, churches, and private schools. The problem lies in attempts to impose creationism on the science curricula of public educational institutions—trying to pass it off to students as science. We appreciate the sincerity of many of the creationists, while rejecting their contention that creationism should be taught as science in public schools.

Because the overwhelming majority of resistance to and organized attacks against evolution come primarily from those associated with Protestant Evangelical Christianity, a corresponding proportion of our discussion will focus on these creationists.[i] They generally believe that Jesus was God on earth, who then died for everyone's sins and rose from the dead. Those who believe this and repent of their sins, accepting Jesus as Lord of their life, are considered Christians, and trust they will go to heaven after death as a result of those beliefs. They seek to lead non-Christians to conversion. Within Christendom, the majority of anti-evolution voices encountered by science instructors are those of biblical literalists (who believe that evolution did not occur and that the universe is less than 10,000 years old). Therefore, though we discuss a wide variety of religious (and nonreligious) rationales for why people are against evolution education, we spend significant time on the literalists' objections to evolution education.

While endeavoring to respect religious culture, we will try to explain the various reasons for creationists' beliefs and what bearing these beliefs may have on students' learning the material on evolution that science instructors teach. The primary purpose of this book, simply put, is to aid science instructors in helping their students learn evolution better.

Over the years, in talking with thousands of high school and college instructors about teaching evolution, we have become increasingly aware that there is a primary core of issues about which teachers inquire. We have written this book to concisely address these issues. We have not

---

[i] Americans' religious preference is broken down as: 58% Protestant, 26% Catholic, 2% Jewish, 1% Orthodox, 1% Mormon. (Moore 2000).

knowingly attempted to dodge any questions, even if they were politically sensitive. Rather, we have decided to bring these questions up and discuss them head-on. After all, students, parents, administrators, co-workers, and school boards often confront teachers with these very questions. While it is certainly true that many of the issues merit entire books, we have attempted to give enough background to familiarize science instructors with the basic issues in order to help them teach better.

It does not help science instructors to hold inaccurate stereotypes about why their students reject evolution and how students feel about this issue. On the contrary, by holding ideas that more accurately reflect their students' rejection and by understanding the sometimes complex culture that supports such rejection, instructors may better address their students' concerns and thus increase learning.

There are many reasons why people fear evolution, and we present the most commonly encountered religious and nonreligious rationales underpinning those misgivings. Also, there are numerous versions of creationism and we will discuss a variety. However, it is too cumbersome to specifically identify each type, each time, and in every context. We will primarily use a working definition for creationism that is related to issues of teaching science in public schools. In cases in which we do not explicitly identify a type of creationism, we will use "creationism" to mean an anti-evolution education position often holding views such as: (1) evolution should not be taught in science courses, (2) evolution instruction should be greatly diminished in science courses, or (3) evolution instruction should be modified to include supernatural cause (or nonsupernatural intelligent designer[s]) as an explanation in science courses instead of, or in addition to, evolutionary theory.[ii] Advocates of creationism will be referred to as "creationists."

When we discuss creationists' actions or views, we are not necessarily referring to all creationists; ascertaining such information would be impossible. Moreover, when we think the flow of the writing would suffer

---

[ii] Many intelligent design (ID) advocates say they would like ID taught without seeking out the identification of the intelligent designer(s)—at least in the science classroom. However, most leading ID advocates (typically, they are Protestant or Catholic) believe that they do know "the" designer or creator.

from unneeded repetition, we will simply state creationists' actions or views. For example, instead of repeatedly stating "90% of creationists do X," "most believe Y," and so forth, we often simply state that "creationists do or believe X or Y."

We will use *evolution* to mean "the descent, with modification, of different lineages from common ancestors. . . . All forms of life, from viruses to redwoods to humans, are related by unbroken chains of descent."[iii]

Many creationists have elevated the conflict between evolution and creationism to the status of a war. In Chapter 1, we describe why creationists consider the conflict so important—as important as many of their fundamental doctrines. With creationist attempts to increase the enlisting of students to carry on battles in science classrooms, the reality is that there are a large number of creationists who consider teachers of evolution "the enemy."

In Chapter 2, we discuss a variety of creationist views, with their differing intensities of resolve, that conflict with evolution and nature of science instruction. An understanding of the students' creationist culture that engenders their misconceptions about evolution in particular and science in general is a major aid to instructional practice.

It comes as no surprise to most instructors that creationist students generally have religious reasons for rejecting evolution. In Chapter 3, we talk about these reasons, how they are engendered, and what happens when creationists perceive that science and their religious beliefs are in conflict. We deal with some of the underlying philosophy of creationism as well. Two characteristics seem to be almost universally present among creationist students: (a) they are pleasantly surprised when they learn that their instructor has some knowledge about their most important beliefs and (b) their admiration and respect for that instructor increases considerably due to this knowledge, which is usually helpful in a teaching milieu. In Chapter 4, we discuss some apparently nonreligious rationales that creationists hold.

[iii] This citation is from a document endorsed by the following scientific societies: American Society of Naturalists, American Behavior Society, American Institute of Biological Sciences, Ecological Society of America, Genetics Society of America, Paleontological Society, Society for Molecular Biology and Evolution, Society for the Study of Evolution, and the Society of Systematic Biologists. (Futuyma 2000, p. 3)

Taking into consideration all the controversy, many creationists and noncreationists alike ask: "Why teach evolution?" In Chapter 5, we discuss some of the many reasons why evolution education is essential.

Creationists often ask instructors creation/evolution questions. We address some of the most typical questions related to science education in Chapter 6, religion in Chapter 7, and general education in Chapter 8. Sometimes science instructors have questions related to how they, as instructors, might proceed. These questions and answers comprise Chapter 9. We focus on pedagogy and teaching suggestions in Chapter 10.

In some sections we try to present various evolution education methods or viewpoints without overtly indicating which ones we favor, thus leaving instructors to decide which might be most proper in principle and/or might work best in their context. Additionally, for ease of reading, we will use "we" even when referring to the actions or thoughts of just one of us. This eliminates the awkwardness of stating repeatedly "one of us has . . ." or "one of us contends that . . ."

Acknowledgments

We thank Stephen Jay Gould for his valuable contribution of time and expertise in writing an insightful Foreword, for reading our manuscript, and for providing helpful comments. We also thank Nadine Ball, Linda Gaither, Charles Granger, Howard Gardner, Ernst Mayr, Anne Ross, Philip Sadler, Eugenie Scott, and Christopher Smith for reviewing our manuscript and providing their comments. Additionally, we thank Graham Bell and Robert Carroll for reviewing sections of the manuscript.

Many persons at Jones and Bartlett helped make this book a reality. Brian McKean understood how important information on creation/evolution was to the academic community and others, and committed Jones and Bartlett to this project. Judy Hauck oversaw the project during its final development and production. Victoria Jones provided day-to-day management, while Mary Hill orchestrated the production process. Anne Spencer provided the exceptional cover design. A heartfelt thanks to everyone on the Jones and Bartlett team, who always do superior, professional work.

Thanks also to Kathy Smith who did outstanding work copyediting the manuscript, and to Nicole Barone and her team at Thompson Steele who provided exceptional production services from manuscript to bound book.

Brian J. Alters
Sandra M. Alters
Montreal
August 2000

I

Creationists Declare War

"There is a war going on in society—a very real battle. . . . but we must wake up to the fact that, at the foundational level, it's really creation versus evolution."

—Ken Ham, Executive Director, Answers in Genesis[1]

"Opposition to evolution is so vocal in the United States that it has threatened federal funding of evolutionary research."

—Nine national scientific societies[2]

"These threats will almost certainly intensify and more forcibly threaten science education."

—Randy Moore, Editor, *The American Biology Teacher*[3]

"Whether or not we like it, we *are* in a battle for the minds of men, especially the minds of our young people, and the enemy is brilliant in persuasion and mighty in influence. Souls by the millions hang in the balance."

—John & Henry Morris, Institute for Creation Research[4]

The topic of the last passage above is about biblical evangelism. Written by two creationist leaders, John and Henry Morris, this passage appeared in a popular creationist book. As is typical of many creationist leaders, the Morrises contend that "the true and whole gospel is founded on creation," and their conception of creation is antithetical to the evolution we teach in science classrooms. They also believe that "evolution and salvation are mutually exclusive concepts."[5] Their battle against teaching evolution is not simply an antiscience fight; they view the teaching of evolution as being against truth, salvation, morality, and the underpinning raison d'être—God. This creationist stance results in an educational reality that is unappealing, but one that science instructors must face. The reality is that there are a large number of creationists who consider teachers of evolution as the enemy.

Part of the difficulty instructors often have with fully appreciating the creationist problem, however, is that many perceive the creation/evolution controversy as analogous to an academic dispute. (Science instructors generally do not view creationist arguments as academically on a par with those for evolution—usually quite the opposite.) However, the adversarial status inherent in this controversy is by no means the "normal" adversarial status we may find in other strong disagreements in academia. The controversy is not, for example, like disagreements among prominent academics who argue for decades in journals, attempting to discredit the rationality of their opponents. Even the most intense and emotionally involved of these and other types of academic controversies do not come close to the emotions and seriousness of resolve felt by many creationist leaders, lay-creationists, and our students concerning the teaching of evolution. Thinking that a creation/evolution controversy in a classroom can be resolved quickly using some standard academic method leads to an underestimation of those differences.

In previous decades, newspaper headlines reported attacks against evolution involving high-profile court cases initiated at the school level concerning control over what public school science instructors must or should teach about origins. These matters generally involved professional and nonprofessional anti-evolutionists attempting to persuade the courts to stop evolution from being taught in public schools or to mandate that some form of creationism be taught alongside evolution in the science classroom. After consistent and significant defeats of these attempts, it appears that many of the professional anti-evolutionary forces are no longer concentrating on this approach. It now appears that they are shifting resources in order to create a war *within* science classrooms.

The intense emotion of the "battle" is now being focused more than ever on enlisting students to engage in fighting evolution education. If one doubts that powerful, professional anti-evolutionists desire a more student-centered front, consider just a few examples from their popular spokesmen (who are rarely, if ever, women). Phillip Johnson's very popular anti-evolution book *Defeating Darwinism by Opening Minds* is a prime example. In it, this Berkeley law professor states who his desired audience is: "In particular, I wanted to write for late teens—high-school juniors and seniors and beginning college undergraduates."[6] He calls his book a "guide" on how to defeat Darwinism. Another book of Johnson's, which has become immensely popular, is *Darwin on Trial*. In it, he characterizes evolution education as a "campaign of indoctrination in the public schools," and emphasizes that something should be done about it.[7] In most of the hundreds of secular bookstores we have visited across North America, whether college bookstores or popular bookstores, we found this anti-evolution book shelved not in the religion section of the store, but in the science section—typically within the evolution books!

On the back cover of another popular creationist student-audience book, the question is posed: "As a Junior or Senior High School student, do you find yourself being hit from all sides with evolution in your public school (and even in some Christian schools!)?"[8] Inside the book students are told the following:

> Think about it like this: Imagine being in a war, and all you know how to do is to throw rocks. Your enemy, on the other hand, has rockets and nuclear bombs. . . . In real-life, the Devil is the enemy of mankind. He wants as many people as possible to be deceived and die without knowing about Jesus and without being forgiven for their sins. That way, he keeps people from going to heaven. Evolution is one of his biggest bombs![9]

There is no other science currently taught in public schools whose opponents use more war metaphors than the subject of evolution. Moreover, no other science has opponents increasingly focused on recruiting students to their cause. The creationist organization Answers in Genesis has helped seed vast numbers of school creationist clubs by providing start-up information and ongoing resource support. Their recruiters have posted the following on their Web site for students:

We want this site to be a resource for you, the frontline soldiers in the
war of truth on the most important battlefield in our society: the
school!![10]

There are plentiful examples of student-centered anti-evolution products,
including books, pamphlets, posters, videos, CDs, and Web courses. We will
touch on some of the more interesting, significant, and influential materials in following chapters.

Along with underestimating the fierceness of emotions involved in the
creation/evolution issue and not realizing that a war is being waged
against evolution in our classrooms, science instructors often hold the misconception that those who share creationist sentiments are a small, disor-
ganized, and underfunded group with minuscule audiences. A common
result of such thoughts is that there is no reason for science instructors to
be concerned about creationists because they and their ideas are harmless.
One point we will make in this chapter is that the contrary notion is true:
Anti-evolutionists are well organized, well funded, and numerous enough
to cause significant harm (and they have). Additionally, anti-evolutionists
have large audiences, and, of most direct importance to science instructors, they believe that they are at war with those who teach evolution.

Campus Apologetics

While it is true that one can find at least a few individuals of almost any
religious faith who reject evolution, the overwhelming majority of resistance to and organized attacks against evolution come primarily from
Protestants associated with Evangelical Christianity. (Evangelicals generally hold that the Bible is the inerrant authority.) Evangelicals reject evolution significantly more than nonevangelical Protestants or Catholics, as
evidenced by their publications and by opinion polls.[11]

Protestant evangelicals believe that they are admonished in the Bible
to be wary of those who attempt to contradict what evangelicals perceive
to be truths revealed in the Bible. Those versed in apologetics (the defense
of the Christian faith) frequently quote from the Letter of Peter concerning this issue: "Be of sober spirit, be on the alert. Your adversary, the
devil, prowls around like a roaring lion, seeking someone to devour."[12]

(5:8). From this and other Bible verses, Christian apologists believe that Scripture directly commands them to be on the lookout for Satan's evildoers.

Josh McDowell is one of America's leading Christian apologists who has spoken to more than 8 million students and faculty in 74 countries. He has been a traveling representative of the Campus Crusade for Christ, an organization that spreads the Protestant Evangelical Christian message on college campuses throughout America. McDowell writes, in one of his books meant to be used as a defense for Christian claims, that:

> Satan and his army of demons desire that we be drawn to the world's standards. . . . As you study the Old Testament, you see that men and women were in a constant struggle with Satan, fighting many spiritual battles. As you study the life of Christ, and Paul, and the other apostles, you see a constant spiritual struggle. Christians today face many spiritual battles.[13]

One aspect to notice is the "us" versus "them" context created by the use of the words "struggle," "fighting," and "battles." If apologists perceive that evolution is in conflict with what is taught in the Bible, then in their words they are involved in a struggle, fight, or battle against the teaching of evolution. The apologist's spiritual battles are not singular to the evolution/creation controversy; it's just one of many that they perceive to be of great importance, and science instructors, whether we like or not, are in the trenches.

Naturally, many people reading the prior passage would say that the spiritual battles recorded in the Bible and those that continue today are totally unrelated to the teaching of evolution. That is, of course, exactly the position taken by most Christian theologians, Bible colleges, and seminaries in the United States, which see no conflict between their religious beliefs and biological evolution. McDowell, immensely popular and considered an authority in the Christian community, disagrees with this interpretation. He responds to one of his most often-asked questions, "Does the Bible allow for the theory of evolution?" in the following manner: "Evolution not only contradicts the Bible, but it also contradicts some basic laws of science."[14]

The leading creationists are essentially apologists who see their primary battlefronts as areas of the natural sciences. They view their ultimate victory as being the rejection of evolution by every student. Because they have been unsuccessful in legislating either the elimination of evolution or the inclusion of creationism into the curricula of science courses in most schools (although they have certainly tried), they strive to educate the students in "science" outside of regular schooling.

For example, the Institute for Creation Research (ICR), a prominent apologetic creationist organization located near San Diego, California, conducts workshops titled "Good Science" throughout the United States for K–6, high school, and adult students. Even by the title of the workshops, students are encouraged to equate creationism with good science and evolution with bad science. Creationists do not use this good–bad distinction primarily to distinguish moral science from immoral science, but rather to contend that creation science has scientific validity, while evolution does not (discussed further in following chapters). In addition to various other presentations, families can go on ICR Creation Tours to the Grand Canyon, Mt. St. Helens, and Yellowstone to be shown firsthand by creationist tour guides the "evidence" for creation and against evolution. Moreover, students' access to creationist educational opportunities and materials has recently greatly expanded with ICR Internet courses, such as the one in the following description:

> [This] twelve-module course (written at the high school level) will permit a student to learn Biblical and scientific creationism at home interactively. . . . Such topics as Creation, the Flood, fossils, and dinosaurs will be presented in Creation Online.[15]

ICR focuses on anti-evolution education primarily through its publications, workshops, lectures, debates, and on-site museum. The key focus of some other evangelical Christian organizations is to spread the word of their Christian faith to others (often referred to as *witnessing*) at institutions of higher learning. One of the largest and best known is InterVarsity Christian Fellowship/USA, an evangelical and interdenominational campus ministry that reportedly had, in one year alone in the late 1990s, over 30,000 students involved in 706 chapters on 534 campuses. They also employed more than 1,000 people, with approximately 650 of their staff working on campuses. They answer the question "Why do you [InterVarsity Christian Fellowship/USA] exist?" in the following manner:

The purpose of InterVarsity Christian Fellowship/USA is to establish and advance at colleges and universities witnessing communities of students and faculty.[16]

They also maintain a presence on college campuses via their publications that they carry in their extensive book catalogs. InterVarsity Press (IVP) states in one of their catalogs that "we believe one way God is leading people into that understanding [i.e., understanding the truth] is through thoughtful, engaging books written, edited and marketed by men and women committed to truth." And the question of "truth" as it relates to evolution is certainly addressed by IVP publications. Although they do not include works by the prominent literalist creationists (see Chapter 2), IVP's catalogs carry and publicize numerous works by anti-evolutionists in their general book catalog for a wide-ranging audience of Christians. Even more interesting is what they carried in their large 1999 "Academic Catalog." This catalog encourages professors (Christian professors presumably) to request complimentary examination copies for books that they might use in their courses.[17] Among the works listed are many by anti-evolutionist Phillip Johnson, such as *Darwin on Trial, Defeating Darwinism by Opening Minds,* and *Reason in the Balance.*

Not only does IVP sell anti-evolution books, they appear to endorse these books and their anti-evolution messages as well. For example, in a critical comments section, the IVP catalog lists numerous accolades, excerpted from various theological journals, of Johnson's *Reason in the Balance.* The catalog prominently displays a quote from the periodical *Preaching:* "One of the most helpful volumes released this year. . . . A must read for thoughtful preachers," and another from the *Journal of the Evangelical Theological Society:* "Darwinian evolutionists have been unable to ignore him [Johnson] and—more significantly—unable to refute him."[18] Also quite telling of the support for anti-Darwinian ideas is the fact that only one other book, out of more than 400 listed in the catalog, has more bulleted accolades than Johnson's book. The catalog also mentions that both *Reason in the Balance* and *Defeating Darwinism* respectively were 1996 and 1998 *Christianity Today* Award winners. In another critical comments section in their catalog, IVP quotes creationist Michael Behe[19] (author of *Darwin's Black Box*): "Phillip Johnson is our age's clearest thinker on the issue of evolution and its impact on society."[20] Moreover, the IVP catalog includes citations telling us that the book "teaches readers to train their 'baloney detectors' on the double think, ad hominems, rhetorical tricks and logical gaps that characterize the public propaganda for Darwinism." These books being praised attack the science of evolution as well as attempt to undermine the teaching of evolution in public schools.

Another large and powerful evangelical Christian organization that is relevant to evolution in the classroom is Focus on the Family. Visiting their Web site (www.family.org) and searching their Resource Center can provide examples of their approach. When we entered the term *evolution* while on the Resource Center page, references to four print materials appeared, three written by anti-evolutionist Johnson and the other titled "Truth Matters" (a collection of folders, notebooks, and posters). Johnson's writings included his anti-evolution education book *Defeating Darwinism* ("suggested donation $10"). Clicking on the title brought up a page showing the cover of Johnson's book with the following remarks written next to it:

> Undo Darwin in your classroom. So many students—and professors— are caught in the controversial debate over evolution. . . . he [Johnson] reveals the key to defeating a bad theory: opening your mind to the truth.[21]

This example shows the intent of the "evolution" resources on this Web site; they are meant to "undo Darwin in your classroom" and are not resources to help students learn evolution.

Focus on the Family also publishes magazines. One of their periodicals titled *Citizen* is described in the following manner on their Web site: "*Citizen* provides an honest appraisal of what's happening in the world news on current events and issues affecting your family that the mainstream media don't always cover."[22] A recent issue included a news item entitled "Darwinists Take Their Agenda to School." The first paragraph begins with the following misinformation:

> Even as mounting evidence casts doubt on the theory of Darwinian evolution—the premise that humans evolved from lower life forms—a major scientific group [National Academy of Sciences] is urging that schoolchildren learn nothing but evolution.[23]

This excerpt has two mistakes: (1) evidence is not casting doubt on evolution, and (2) the National Academy of Sciences (NAS) is *not* urging that only evolutionary explanations be presented in schools, only that science explanations be included in science courses and nonscience explanations be included in nonscience courses. The news brief then continues by citing anti-evolutionist Phillip Johnson:

> The scientific and educational elites are really baffled that the public doesn't believe them after they've drenched us with propaganda for decades. . . . They're frustrated.

Science instructors should not underestimate such organizations' influence on students' science learning. Fellow creationist students, friends, parents, and others may encourage students to seek out and use resources from Web sites and periodicals produced by such organizations, and students may hear about these Web sites and periodicals on radio programs. This could be significant, given that an estimated 25 million Americans are turning to Web sites for religious purposes.[24] Focus on the Family broadcasts daily from more than 2,900 North American radio facilities.[25]

Some evangelical Christian publications are designed for teachers. A popular Focus on the Family periodical is *Teachers in Focus*, which is intended specifically for educators to read, and read it they do—over 30,000 educators subscribe.[26] Compare that to the approximately 10,000 subscribers of *The American Biology Teacher,* a life science education journal having the largest circulation in the world (and that appropriately supports the teaching of evolution in the science classroom). By comparing these circulation numbers, one can begin to see the potential that evangelical periodicals such as *Teachers in Focus* have for expanding the numbers of anti-evolution educators and therefore anti-evolution students. In an issue of *Teachers in Focus,* the editor authored an article titled "Skewed Science," in which he severely criticized a recent evolution education publication by the National Academy of Sciences.[27] He states that "Evolution can readily account for changes in bird beaks. But new organs and body plans are a different matter." He proclaims that "teachers would have to teach a skewed science curriculum that misrepresents broad areas of biology."[28] In this same article, the *Teachers in Focus* editor supported a book written by creationist Michael Behe.

This section presented just a few examples of actions to spread the word about science instructors' "false" teachings of evolution. We have only scratched the surface here.

The War Chest

Begun in 1994, the Answers in Genesis creationist organization (AIG) currently enjoys an approximate $5 million annual budget, a staff of 55 people, a 100,000-person mailing list, and headquarters on 47 acres in Florence, Kentucky. In one year alone, AIG speakers have reached over 150,000 people at their meetings. AIG Executive Director Ken Ham hosts a daily radio program heard on 260 stations in America and 40 overseas, and receives over 400 speaking invitations annually. In 2001 they report "averaging close to 11,000 people (not 'hits' but individuals) per day" visiting their Web site. In the year 2000, they had over 2.5 million visitors to the site. AIG is currently constructing an $8 million 95,000-square-foot building to house a creationist museum and reports that it will have the largest collection of life-sized dinosaur models in America.[29] Why a museum?

> This museum will be a wonderful alternative to the evolutionary Natural History museums that Satan has used to influence so many minds.[30]

The most popular creationist organization is the Institute for Creation Research (ICR) mentioned in the previous section. With an approximate $5 million annual budget, they own and operate from a 21,000 square-foot building, and a new and much larger building next door.[31]

> We believe God has raised up ICR to spearhead Biblical Christianity's defense against the godless dogma evolutionary humanism. . . . by showing the scientific bankruptcy of evolution.[32]

ICR reports having reached millions of people. It is difficult to estimate the ultimate effect such an organization has on students to challenge evolution, but its production is noteworthy. ICR has conducted over 300 debates, usually held on university campuses, with audiences mainly of students and faculty and attendances of up to 5,000. They have given anti-evolution seminars, lectures, and workshops in thousands of churches and schools in every state and in over 40 countries. One of their newest outreach endeavors for high school and college students is a two-day, 14-hour workshop that is a "creation-equipping course designed to train students in creation apologetics. . . . [assisting those] who lack confidence to challenge the flawed claims of evolution in the classroom."[33] Additionally, ICR personnel have written and published over 100 books; the organization carries over 150 titles, including periodicals, videotapes (over 50), CDs, and pamphlets. Some of their periodicals have a circulation of over 300,000 per month.

ICR has more than 50 full-time employees, and their anti-evolutionary messages have been broadcast via a radio network through some 1,500 outlets. They have an accredited graduate school, which offers M.S. degree programs in the fields of Astro/Geophysics, Biology, Geology, and Science Education. Their Museum of Creation and Earth History reports an annual attendance of adults, children, and school groups numbering almost 25,000 per year. And students' access to their views and their educational materials is expanding via their online courses; they receive over a million "hits" monthly on their Web home page.[34]

Quickly after the Kansas State Board of Education removed virtually all mention of evolution in the state's curriculum in 1999, ICR published a small book about the event. One of the chapters, "Principles and Guidelines for Action in your State," outlines how creationists can get involved locally. This chapter ends with the following three sentences:

> Be prepared, for the fight will be long. The other side may not be fair, and you will lose some of the skirmishes. But, it is a worthy fight, and one that will give you great satisfaction.[35]

One of the newest creationist organizations is the Center for the Renewal of Science and Culture, which is organized under the umbrella of the Discovery Institute, headquartered in Seattle, Washington. They feature the books by Phillip Johnson and Michael Behe mentioned previously. In July 2000, the organization held a national symposium titled "Darwin, Design & Democracy: Teaching the Evidence in Science Education." An anticipated highlight of the program included presentations on "practical implementation of design theory in science education."[36] (Design theorists hold that evolution is an insufficient explanation for the diversity of life on earth, and believe that science should recognize an intelligent designer, whether supernatural or nonsupernatural extraterrestrial. This idea is discussed in Chapter 2.)

Such intense involvement of local students and adults in the anti-evolution battle is primarily an operation conducted in the United States. However, creationist operations have already traveled far, having grown in popularity in Europe, Asia, and the South Pacific.[37] Various types of creationist organizations already exist in many other countries; one of the oldest is the Biblical Creation Society in Great Britain. In Quebec, Canada, where 90% of the population speaks French, there is a popular bilingual (i.e., French/English) creationist organization: The Creation Science Association of Quebec/L'Association de Science Créationniste du Québec. Since 1984, the founder of this organization reportedly has been invited to speak at more than 400 church congregations, many colleges and universities, and some Christian and secular schools.[38]

There are scores of creationist organizations worldwide (creationists believe that there are probably over 100), each with their own sources of funding, and they vary greatly on what they believe about creation.[39] Some are connected directly to specific religious denominations, others have more general religious affiliations, and a few are completely independent. But the combined contents of the creationist war chest—in terms of personnel, funding, and products—are significant. Moreover, these creationist organizations with their significant resources have a common desire: to diminish greatly the explanatory power of evolution or to extinguish it completely in public school science classrooms.

The Motivation for Attacking

Let me issue a call to theologians, pastors, and Christian leaders. Christianity is engaged in a worldview war and needs all her soldiers.

—John Morris, President, ICR[40]

What is the "worldview war" in which Christianity is engaged? Almost all literalist creationist apologist leaders contend that evolution is diametrically opposed to the version of events given in the Bible. These leaders are convinced that the Bible indicates clearly that the diversity of life on earth is *not* a product of evolution, regardless of whether God controlled the evolution. (We elaborate further on the topic of theistic evolution in Chapter 2.) They understand the Bible to plainly report that God created Adam and Eve in pretty much the same form as humans exist today; they did not evolve from any lower forms of life. (Recent Gallup polls report that 33% of American adults believe that "the Bible is the actual word of God and is to be taken literally, word for word.")[41]

These creationist leaders also believe that the Bible is the one-and-only truth, and when they read accounts in the Bible, they read them as historical truths, rarely as metaphors. To them, the narratives in the Bible are not the same as those in any other books ever known to exist. The Biblical records report the most important aspects of humans' lives—where we came from, why we are here, and where we are going after we die. The biblical records also tell us how we should live our lives, how we should view the laws of the land, and what our relationships should be with our parents, spouses, children, and non-relatives. Many noncreationist Christians, of course, think that there are many truths to be learned and believed from the pages of the Bible. But peculiar to the creationists is their fervent belief that humans are not the product of evolution—a belief held as deeply (or nearly so) as their conviction that Jesus was born of the Virgin Mary, performed miracles while living on earth, arose from the dead, and is now part of the Trinity (Father, Son, and Holy Ghost).

The emotional ties to these beliefs cannot be overemphasized. These beliefs have to do with knowing that the Creator of the universe (the heavenly Father) loves them, and that there are absolutes. If they successfully pass God's judgment, there will be a pleasant life after death and they will potentially see loved ones again who have died (providing the loved ones successfully passed judgment also). To the creationists, the accuracy of creationism is fundamental to many, if not all, of these beliefs.

If evolution is right, if the earth is old, if fossils date from before man's sin, then Christianity is wrong!

—John Morris, President, ICR[42]

It is often difficult for science instructors who do not share creationist beliefs to empathize with the strong feelings of students who hold such beliefs, which are foundational to their worldview. An inability to empathize with these feelings is one of the reasons why many science instructors have trouble understanding that creationist students feel a strong moral obligation to fight against the teaching of evolution, in addition to their having incredibly strong cultural reasons to abhor it. Creationist students' feelings on this issue can be thought of as analogous to the way a science instructor might feel if someone were teaching students something they thought was blatantly reprehensible. Examples of what might engender the greatest emotional response in a science teacher would vary greatly among individuals, but could include teaching students great falsehoods, such as one race or gender being inferior to another, the Holocaust never having occurred, or people with AIDS deserving to die. All instructors whom we know would rightfully do their best to combat such outlandish teachings. And this is precisely the point: Many creationists feel even stronger emotions about those who, in their view, teach students such great falsehoods as evolution, particularly what they consider to be godless evolution.

This idea that evolution is a great falsehood (and evil) is clearly seen in the writings of Henry Morris, the most prolific creationist author and probably the most influential creationist of the latter half of the twentieth century. He is President Emeritus of ICR, former Head of the Civil Engineering Department at Virginia Tech, and author of over 45 books on biblical and scientific creationism. He writes that

> The doctrine of special creation then pervades all the rest of the Bible and is the real foundation of all other truth, especially including all the great doctrines of the Christian faith. The evolutionary concept of history, on the other hand, is Satan's greatest weapon in his long war against God, serving him as the root of every false philosophy and evil practice known to man.[43]

This statement was not taken from a typical creationist book but rather from a "study Bible" annotated by Morris titled *The Defender's Study Bible.* "Study Bibles" are popular in Christian culture. These texts are similar to other Bibles and contain the entire text of the Bible, but they also have the important addition of commentary, usually from a well-known Christian leader or theologian. (Study Bibles are discussed further in Chapter 3.) Notice the words in the prior quoted passage that are used in relationship to evolution: "Satan's greatest weapon" and "war against God." As we noted previously in this chapter, the militaristic symbolism used here is not peculiar to this one popular creationist organization or to one or a few creationist authors; the overwhelming majority of creationist books and other products we have encountered set up this adversarial condition.[44]

Emotional Connectedness

Most educators know that student learning involves a great many factors, one of which is students' emotional connectedness to an issue. For many people, evolution is far less emotionally satisfying than creation. It is certainly easier for evangelical Christian leaders to structure situations that foster emotional connectedness to the creation issue than it is for science teachers to create situations in the classroom in which students come to feel "emotionally connected" to the topic of evolution.

Consider, for example, how incredibly motivating songs are to many people, especially young people. The great patriotic songs of the United States have been used to encourage young Americans to defend their country in times of war and to inspire Americans of all ages to feel that "our side" is more virtuous than the enemy. The words of such songs, combined with the music, provide a powerful emotional tug that often can motivate individuals far more than mere words alone. The effect of growing up singing hymns (or at least listening to hymns) at least once a week for many years can certainly make a significant impression.

In addition to such emotional factors, it is not uncommon to hear conservative Christian church leaders tell their congregations that the hymn they are about to sing or just sang is a sermon in itself. (To the disappointment of many a church-going child with a little too much energy, this short sermon-in-a-hymn assertion never prevented any pastor from preaching the sermon of the day.) These pastors know how powerful the message sometimes is on the pages of the hymnal and are quick to point it out. In many cases the words are authored by well-known church leaders of the past, those of great prominence in the church's history. The authors of these words often are of much greater stature in the church than the head pastor who will be delivering the sermon. In some cases, the words are taken directly from verses or entire chapters of the Bible. The primary purpose of the congregation singing hymns in conservative Christian churches is not for entertainment but to fulfill scriptural instruction to give God praise through song. But most church leaders will quickly point out that since most hymns contain or reference God's Truth, the hymns also can minister to the singing and listening congregation as well.

To illustrate, one hymn sung traditionally in most creationist congregations is "Onward Christian Soldiers." Even many non-Christians have heard this religious song. It is thought by many pastors to help their congregation with problems concerning conflict and courage. Not only have Christian elementary and junior high school principals had their choir students sing this hymn, but also they expect all students in schools with required weekly chapel services to sing along, as would a congregation in church. We are not referring to the new kinder and gentler version of this song (more acceptable to pacifistically oriented theologies), which some noncreationist churches have adopted rather recently, but to the traditional version that is still sung in most, if not all, creationist churches (and probably most other conservative churches as well). The first couple stanzas set up how Christians move (or should move) against Satan's legions.

> Onward! Christian soldiers,
> Marching as to war,
> With the Cross of Jesus
> Going on before.
> Christ, the royal Master,
> Leads against the foe;
> Forward into battle,
> See! His banners go.
>
> At the sign of triumph
> Satan's legions flee;
> On then, Christian soldiers,
> On to victory!
> Hell's foundations quiver
> At the shout of praise;

Brothers, lift your voices,
Loud your anthems raise

Like a mighty army
Moves the Church of God

We realize that it is almost inconceivable for many science instructors to comprehend that when many creationists sing this hymn they envision that one of the foes that needs to be beaten in today's society is teachers of evolution, especially those who are nontheistic evolutionists. (Non-Christians who teach evolution generally get a much harsher treatment than Christians who teach an evolution that is guided by God.) However shocking, this is the situation for many, and it contributes to making the struggle emotional as well as academic. Evolution is perceived and felt as being an omnipresent primary evil.

> Through the education system . . . evolutionary thinking is all pervasive. Creation versus evolution is NOT a side issue. They are really the front lines of the battle.
>
> —Ken Ham, Executive Director, Answers in Genesis[45]

The president of ICR understands the importance of hymns and has recently authored a 250-page book of song and scripture. In it he writes "Many times scripture compares this life to a battle, with the Christian a soldier warring against the foe."[46] The final three pages of the book are advertisements for creationist books for adults and children.

When conservative Christian church leaders give sermons concerning the Protestant hero Martin Luther and his stand on Scripture alone as the arbitrator for truth, they often will use this opportunity to illustrate how a great historic church leader stood his ground, claiming the accuracy and authority of the Bible above all else, including going against those leaders of the church who make proclamations contrary (in Luther's perception) to the Bible. Henry Morris and other creationists keep the following statement by Luther in the flyleaf of their Bibles:

> If I profess with the loudest voice and clearest expression every portion of the truth of God except precisely that little point which the world

and the devil are at that moment attacking, I am not confessing Christ, however boldly I may be professing Christ. Where the battle rages, there the loyalty of the soldier is proved, and to be steady on all the battlefield besides, is mere flight and disgrace, if he flinches at that point.[47]

Church leaders will typically encourage their congregations to do as this passage suggests in the non-Christian world. Such sermons are usually preceded or followed by the singing of Martin Luther's legendary composition "A Mighty Fortress is Our God" (1529), which tells that "though this world, with devils filled, should threaten to undo us, we will not fear, for God has willed His truth to triumph through us." Many perceive this hymn as setting up adversarial roles—believers versus unbelievers—by using terminology such as "our side" and "the battle." The final stanza talks of the ultimate sacrifices that Christians should be willing to endure to uphold the truth: "We answer his commanding. Let goods and kindred go, This mortal life also; The body they may kill: God's truth is ruling still."

For most Christians, conservative or not, these hymns have absolutely nothing to do with science instructors who teach students about evolution. In fact, most Christians would probably find such a proposition to be so irrelevant as to be laughable or would find it sad that other Christians have construed some hymns to have such meaning.[i] But for those who are raised in a creationist culture where evolution is considered an enormous evil that must be fought, hymns such as this are perceived as applicable to such spiritual battles and are explained as such. To this point, John Morris recounts that the most powerful rendition he ever heard of a Luther song was at a recent ICR Graduate School graduation ceremony.[48] There are many spiritual wars that Christians feel they must fight, and hymns about fighting for what is right, about courage and conflict, are often applied to the battle at hand.

So, in the classroom when science instructors present evolution, they are often presenting not only an academic challenge to students' misconceptions, but also a great emotional challenge to creationist ideas that have

---

[i] Even though many creationists are motivated *indirectly* by singing Luther's words to battle against those who would teach evolution, most lay-creationists may be surprised that Luther did speak explicitly about literalist creation. Luther wrote: "We assert that Moses spoke in the literal sense, not allegorically or figuratively, i.e., that the world, with all its creatures, was created within six days, as the words read." (Pelikan 1958, p. 5)

been engendered through various sources, including the emotionally powerful medium of song. While there are a few comical songs concerning evolution that instructors can use in the classroom to amuse their students, such songs will never engender the great passion some students may feel about their Creator.

The Unbelievable Battle

Most creationists are quick to point out that some practicing scientists are creationists. And it is true that a very small number of practicing scientists find evolution scientifically untenable. Surprisingly, a few even hold positions at respectable research universities. The primary purpose for creationists bringing this fact to noncreationists' attention is an attempt to gain scientific credibility by noting that they have scientists on "their side" too. However, scientists who have written papers about why evolution is scientifically untenable have not been able to publish their ideas in standard scientific journals. When creationists learned that these ideas, in the past few decades, have not been able to pass scientific muster via peer review, the creationists respond that many of their creationist colleagues *have* been published in reputable scientific journals.

This is true; articles by creationists have been published in standard scientific journals in areas such as biochemistry, biology, and physics, but their papers do not contain biochemical, biological, or physical data or arguments that attempt to counter evolution. Creationists seem to think, however, that if someone publishes an article in a standard scientific journal concerning human anatomy (containing no explicit anti-evolution relevance), then they have genetically justifiable arguments and data about why evolution is impossible. If someone publishes an article concerning applied hydraulics in engineering (containing no explicit anti-evolution relevance), then they automatically possess geologically justifiable arguments and data about why the earth is only 10,000 years old. These contentions are obviously unsound.

If this creationist argument is countered with the idea that these publications have nothing *explicitly* to do with challenging evolution, they usually agree. Creationists contend that they conduct research, the results of which challenge evolution, but claim that they cannot get this work published in standard scientific journals because of anti-creationist bigotry that exists in the scientific community . . . and the scientific community controls science journals. Furthermore, they assert, the reason their anti-evolution papers are rejected is *not* due to the fact that their research is flawed.[49]

A counter to these arguments includes a history lesson. Hundreds of years ago, publications in science were much more creationistic than evolutionary. As time passed, scientific publications became less and less creationistic and more evolutionary, arriving at the point today where it is impossible to find standard scientific journal articles attempting to challenge evolution. The reason for such dramatic and diametrical change is twofold: (1) generations of scientists have been compelled by the overwhelming evidence to conclude that evolution is scientifically tenable and creationism is not, and (2) the scientific community gradually, over the centuries, increased its commitment to methodological naturalism as a fundamental principle of science.[50] (Methodological naturalism means that scientists use methods that pursue natural causes of phenomena rather than supernatural causes. We discuss this topic further in Chapter 2.) The response by creationist leaders and their followers to these assertions may be shocking to practically all science instructors: They openly allege that evolution did not gain its status as the scientific theory for life's diversity through rational scientific exploration of the data over the years, but rather that evolution has become fundamental to the life sciences for religious reasons. As outrageous as this may sound, these creationist leaders believe that the rise of evolutionary theory and the decline of creationist convictions in science is primarily the result of one long war waged against God by the scientific community! Even Darwin does not escape such accusations; Henry Morris writes that

> Darwin, however, wanted to find a way to escape Paley's conclusion [that organisms have irreducible complexity and therefore have been intelligently designed], not for scientific reasons, but because he refused to accept a God who would condemn unbelievers like his father to hell.[51]

Concerning today's scientists, Morris believes

> the evolutionary establishment in science or education . . . . hold their position for religious reasons, not scientific.[52]

Creationist leaders are not going so far as to claim that somehow scientists are all involved in a massive interdisciplinary conspiracy to overthrow creationist dogma. Rather, they claim that somehow scientists were more likely than others to accept a naturalistic worldview—one in which natural forces cause things to occur. Creationists believe that holding a naturalistic worldview is a sin because naturalism removes God from various activities, such as specially creating the planets, stars, and all organisms including humans (i.e., Adam and Eve) in pretty much the form we see them today. They are suggesting that competent scientists only come to the conclusion that evolution is the superior scientific theory because they possess naturalistic worldviews. Creationists see this dichotomy as an uncompromising conflict between two worldviews that both make claims concerning the history of life.

> These [evolutionary] ideas destroy the foundation for the Gospel and negate the work of Christ on the cross. Evolution and salvation are mutually exclusive concepts.[53]

> —John Morris, ICR President

The primary error in this sort of contention is that many scientists who have concluded that the diversity of life is the result of evolution also believe in God, many being devoutly religious Christians.[54] The results of polls of scientists in both 1916 and 1996 indicate that roughly 40% of scientists believe in a personal God.[55] This percentage is consistent whether the scientists were practicing at the beginning or at the end of this century.

Obviously, these scientists find no conflict between their scientific work and their religious beliefs. They have a wide variety of beliefs concerning evolution and God, ranging from believing that God controls every step of the evolutionary process (although they admit that this control is undetectable by scientific inquiry) to believing a doctrine that God uses the true randomness of evolution as the mechanism of choice. The literalist creationists contend that such scientists, although Christians, are falling short of Christian standards for proclaiming that evolution is accurate.

Anti-Evolution Education Evangelism

What do the creationists think about the millions of students who are not taught creationism in public science curricula and as a result accept evolution as the scientifically acceptable theory? They think that this is a horrible tragedy, and that the solution is to evangelize these students and instructors. The Morrises write about the "stranglehold" that public institutions have concerning evolution education:

> The public school is essentially a mission field in which the gospel needs to be preached . . . . scientific creationism can well be regarded as the cutting edge of the sword of the Spirit with which we are to preach the gospel.[56]

> How are they to be reached for Christ? . . . It would seem obvious that biblical evangelism—which especially includes *creation evangelism* [emphasis added] in most such cases—must be central in the methods employed by concerned Christians and churches . . . . Christians are, indeed, involved in warfare.[57]

In light of these ideas, then, it is not difficult to understand why many devoutly religious Christian creationist science students feel they are soldiers in an all-important war, particularly if they believe the Morrises' writings. (The Morrises are probably the most influential creationists living today.) Science instructors who see no conflict between teaching evolution and holding their religious beliefs, but who are not very knowledgeable about creationist theology, might be appalled to learn what "side" creationists perceive them to be on in their war. The Morrises write:

> We [creationists] are on the winning side in the battle for the mind, as surely as the world exists! Satan has "blinded" many minds (2 Cor. 4:4) for a little while, but one day "the devil that deceived them" will be "cast into the lake fire" (Rev. 20:10). In confidence of God's truth and in assurance of that day, we even now are active in the battle.[58]

Here again we see creationists engender the war hyperbole and symbolism of the controversy. Recently ICR published and distributed an article to its members entitled *A Call to Arms for Conservative Christian Science Educators,* which states: "The evidence is indisputable, and the battle lines are drawn."[59]

Some ICR books also focus on the "war." In an ICR pamphlet that advertises anti-evolution books, the cover of *Weapons of Our Warfare*, which displays two large criss-crossed broad-edged swords, is shown.[60] Adjacent to this cover is another showing students in a classroom facing a teacher who is pointing to a chalkboard with the word *evolution* written on it. The title of the book is *Someone's Making a Monkey Out of You*.[61] The implication of this title is that instructors who are teaching evolution as scientific fact are trying to make fools of students. These books are only two examples of many.

ICR produces other print materials with the same anti-evolution "war" focus. For example, ICR sells a poster that has a cartoon drawing of two old castles, each built on its own small island. One of the castles is built on an island with "Creation" written across its foundation, a "Christianity" flag waving, and clergy firing military cannons at the other castle. The other castle is built on an island with "evolution" written across its foundation, a "Humanism" flag waving, and pirates firing only one cannon at the Christian castle. The humanism castle is in perfect condition. However, the Christian castle is starting to fall apart because its creation foundation is being hit by cannon fire from the humanism castle's only cannon. The pirates in the humanism castle appear to be so confident in their undermining of the opposing castle's creation foundation that they are smiling, laughing, and blowing up brightly colored balloons that have written on them "no absolutes," "homosexuality," "promiscuity," "abortion," "lawlessness," "nuclear arms," "pornography," "drugs," "divorce," "immorality," "euthanasia," and "racism." This image is consistent with much of the creationist beliefs that not only is evolution contrary to what is taught in the Bible but also that evolutionary theory is foundational to what creationists perceive as many of today's social ills. Additionally, the symbolism is clearly warlike.

Unfortunately, many science instructors who don't wish to enlist in any battles find themselves on the front lines. Furthermore, creationist students they encounter in the classroom do not necessarily see the conflict as being just between them and the instructor. These students see the conflict as a battle to stop the instructor from spreading false, damaging, and evil concepts to others, especially to other to students. Creationist leaders repeatedly admonish their followers to take the battle to the educational institutions:

Christian creationists should do all that they can to inform, encourage, and persuade school boards, school administrators, and teachers to teach scientific creationism as at least a scientific alternative to evolution. The same is true of college and university administrators and faculty members.[62]

Most creationist leaders believe that the inciter of sin is the Devil, that mankind should resist the Devil's lies, and that one of the most powerful lies the Devil is currently employing is evolution. Certainly no creationists we know think that science instructors are demons. But many do believe that science instructors are unintentionally aiding the Devil by spreading the lie of evolution. So if the creationist student perceives that the teacher is simply not aware of the truth of creationism, the teacher might be subject to creation science and theology lessons offered by the student in the science instructor's office or during class. However, if time passes and the student thinks that the instructor is well aware of the evil implications for teaching evolution, or, worse yet, if the student perceives that the instructor is well aware of the "implications" and *welcomes* them, many students will consider that science instructor as an enemy—an enemy to themselves, creationism, Christianity, Jesus, and God—to name a few. Whether we like it or not, science instructors are an integral part of the creationists' war and that war is affecting the teaching and learning of science.

We may laugh at a marginal movement like young-earth creationism, but only at our peril—for history features the principle that risible stalking horses, if unchecked at the starting gate, often grow into powerful champions of darkness.[63]

—Stephen Jay Gould, Harvard University

2

Creationist Students' Culture and Nature of Science

"The stifling stranglehold that Darwinism
has exerted over our educational system must be loosened."

—Duane Gish, author of
Teaching Creation Science in Public Schools[1]

"Evolution was promoted as science,
but it is not science—it is a belief system about the past."

—Ken Ham, Executive Director,
Answers in Genesis[2]

Most educators would probably agree that it is important to know why students think something they are being taught is inaccurate. Yet when it comes to students rejecting their teaching of evolution, many educators just chalk it up to students being creationists and do not explore their reasons any further. However, the label *creationist,* while often useful for categorizing the wide variety of people who reject evolution, is much too broad to give educators an appropriate understanding of the numerous rationales students have for rejecting the underlying theory of biology.

In Charles Darwin's time, the "creationist" label was generally used to refer to someone who believed that the human soul was not inherited from the parents but was a special creation for each individual. However, the day after *The Origin of Species* became public, Darwin began writing letters using the term *creationist* to refer to anti-evolutionists.[3] The term as it is used today has come to mean specific types of evolution rejection, which vary greatly depending on what you read or with whom you talk.

For example, many science instructors believe that anyone who rejects evolution must be a religious literalist fundamentalist and/or someone with a conservative political agenda. However, polls show that about half of Americans choose options other than evolution to explain how humans arose on earth. These figures indicate that more persons than just religious fundamentalists (let alone literalist fundamentalists) or political conservatives choose nonevolutionary options. A Gallup poll reports that about 56% of conservatives, 42% of moderates, and 36% of liberals choose the survey option "God created human beings pretty much in their present form at one time within the last 10,000 years or so."[4, i] Gallup also reports that about half of Republicans and half of Democrats choose this view as well, leading us to believe that the rejection of evolution is bipartisan.

Many students who reject evolution *do* have rationales for their objections. Some of these rationales are well thought out, while others border on the affective domain—responses that stem from emotion. The cognitive rationales range from what most people would consider to be purely religious rationales to rationales that may strike many as nonreligious. The vast majority of students, however, hold some combination of religious and nonreligious rationales for their rejections.

Instructors should be aware of students' conceptions in order to help them learn the science of evolution better and to understand why the scientific community agrees that evolution is the only scientific theory to explain the diversity of life. Otherwise, it will be difficult, if not impossible, to productively address students' misconceptions about evolution. Additionally, to better understand why many students (and nonstudents)

---

[i] The percentage who chose this nonevolution response is also similar among regions: East 40%, Midwest 44%, West 48%, South 54%.

contend that the evolutionary science we teach is inaccurate, it is illustrative to examine some of the religious and nonreligious rationales underpinning their thinking. In the following chapters we will look at some of these specific yet greatly varied religious and nonreligious rationales that students typically give for their rejection of evolution.

The vast majority of student rationales for rejecting evolution fall outside the context of the public school curricula. Therefore, these conceptions about evolution are most likely engendered through nonformal learning activities. The source of anti-evolutionary information for most of these activities ultimately comes from educated evangelical adults who believe (primarily for underlying religious reasons) that evolution never occurred. Evidence of widespread anti-evolution evangelical interest sometimes comes from unexpected sources.

One excellent example of the evangelical interest in anti-evolution is the journalistic masthead of evangelical Christianity: the periodical *Christianity Today*. Its Founder and Chairman of the Board of Directors is Billy Graham. In the late 1990s, this popular evangelist drew an audience of 283,000 in Florida for a single sermon. Another of his sermons reached an audience of one billion via satellite.[5] He has said that he considers *Christianity Today* "a rallying point" for evangelicals, "a flag . . . under which we all can gather," a "strong vigorous voice to call us together," to "reach the clergy and laity of every denomination."[6]

Christianity Today is not a fundamentalist-literalist periodical. Since its readership spans the continuum of evangelical theology, it enjoys one of the highest circulations of all Christian magazines. Given this large and theologically varied evangelical readership base, it would be reasonable to expect that one of *Christianity Today's* recent books of the year would be about subjects such as witnessing (proselytizing), how to get to heaven, or how to live a better Christian life. However, out of the more than 200 books nominated, then reduced to a list of 26 titles by "a large panel of scholars, pastors, writers, and other church leaders," the book of the year selected was *Darwin's Black Box: The Biochemical Challenge to Evolution,* authored by a biochemist.[7] Interestingly, 25 years prior to this award *Christianity Today* selected as one of its "choice evangelical books" of the year *The Genesis Flood*, authored by John Whitcomb and Henry Morris, the latter being founder and President Emeritus of the Institute for Creation Research (ICR).[8] In this same issue, *Christianity Today* also contains full-page advertisements for Phillip Johnson's *Defeating Darwinism by Opening Minds*. The advertisements state that Johnson, a law professor at Berkeley, "shows how ordinary Christians can defeat the false claims of Darwinism."[9]

So, contrary to the popular characterization that anti-evolutionism is relegated to solely fundamentalist-literalist concerns, it appears that it is an important theme for the larger evangelical Christian community. Therefore, categorizing the myriad evangelical views for rejecting evolution becomes a difficult, if not impossible, task. However, we use the term *creationism* in a broad sense throughout this book and provide, in the next section of this chapter, easy-to-understand demarcations of its various forms. We hope this will clarify the meaning of the term *creationism* and eliminate some of the ambiguity that typically surrounds it.

Before we begin, some caveats: Designation almost always draws criticism from various camps. Such is the nature of attempting to demarcate anything that has great importance to people, especially to those who either proudly go by the term being characterized or vehemently do not want to be labeled by the term. We contend that increased clarity is inherently important for instructors' understanding of the culture of their creationist students. So we will proceed regardless of the probable onslaught of criticisms that we might receive for purportedly not incorporating some particular theological view, causing the production of some synergism of two theological views, or misrepresenting a particular denomination or faction.

Most creationist views can be represented within the categories of literalist, progressive, theist, and intelligent design.[10] The names of these categories are often used in other contexts and the particular demarcations we draw here are not used universally. In fact, some who hold positions within these categories would not even consider themselves creationists, thus illustrating further the need for greater clarification.

Clarification is also needed as a matter of general education, because only 53% of Americans recall having heard the term *creationism*. (Of those, 80% say they are either "very familiar" or "somewhat familiar" with creationism.)[11] Additionally, we hope that categorizing the different types of creationisms will facilitate discussions among science instructors about evolution/creation issues and will help in understanding the positions held by the various creationists they may encounter.

Perceived Biblical Challenges to Evolution Teaching

Literalists

Christian literalists believe that the Bible does not just contain the word of God; it *is* the word of God.[ii] This means that *every* portion of the Bible is considered God's revelation to mankind. Literalists believe that what is written in the Bible is divinely inspired, and that while humans may have done the job of physically writing down most of the words (the Ten Commandments being physically written by God himself), each and every word as recorded in the original texts was inspired by God. Though most Christian leaders would agree that the originals are long gone, they have a high confidence level that the Bibles we have today are accurate reproductions of original texts.

Unlike other types of creationists, literalists believe that the Bible is an authority on all matters on which it speaks—not just issues concerning ethics, morality, or theology. However, literalists disagree on which matters the Bible does, in fact, speak. Nevertheless, almost all literalists agree that biblical stories are historical facts and that biblical reporting is unquestionable. Some of the biblical stories that literalists consider historical facts are: Adam and Eve being the first humans created directly from the dust of the ground, Eve being formed from Adam's rib, Noah's flood covering the earth, Noah's ark saving two of every kind of terrestrial animal, Jonah living for several days in the belly of a great fish, Jesus being born of the Virgin Mary, Jesus performing miracles (e.g., walking on water, turning water into wine, healing the sick, raising the dead), Jesus being resurrected from the dead, and Methuselah living 969 years.

The Bible is also considered to be infallible by the literalists. They believe that when the Bible gives advice, instructions, warnings, and so forth, there is no possibility that the advice might turn out to be bad, some of the instructions faulty, or warnings unnecessary.

When it comes to science, just as with other matters, literalists believe that the Bible is infallible, inerrant, and authoritative. Genesis is read literally. Therefore, literalists believe that each day of the Genesis creation

---

[ii] A recent national poll (Gallup News Service 2000) reports that 33% of respondents chose "the Bible is the actual word of God and is to be taken literally, word for word."

account is a natural solar day. Because of this literal reading, they do not believe that the diversity of life on earth is due to evolution but rather that the first two humans—Adam and Eve—and all the other organisms were suddenly created within four 24-hour days in the first week when God created planets, stars, and light.

Most literalists also accept a chronological record of Genesis, contending that the creation of those first organisms and the universe took place approximately 10,000 years ago. Thus, literalists are commonly called *young-earth creationists*. It is out of this theological heritage that most of the leading anti-evolutionary organizations arise, the most popular, and probably most influential, being the Institute for Creation Research (ICR). ICR's tenets of biblical creationism state that

> All theories of origins or development which involve evolution in any form are false.[12]

This one tenet distinguishes literalist creationism from other types of creationism. But as we describe in Chapter 4, literalists actually do accept some "kinds" of evolution—it is all a matter of how they define it.

Progressives

As compared to the literalists, progressives are more widely dispersed in their beliefs but share the view that the earth is much older than what literalists contend. One faction of progressives holds a view that is quite often referred to as *gap* or *ruin and restoration* interpretation. This group of progressives believes that earth was supernaturally created as described in Genesis 1:1 (including life on earth even though it is not explicitly stated in the verse), that this life eventually became extinct, and that other life was recreated in Genesis 1:2 and after. The duration of the gap is generally considered to be indeterminate—possibly billions of years. Therefore, the gap accounts for the old age of the earth and fossils, yet advocates still retain a belief in a recent supernatural creation of Adam and Eve and all current life on earth.

Another camp of progressives adhere to a *day-age* interpretation of Genesis and believe that the days of creation listed there should be interpreted as lasting indeterminate periods of time—even millions or billions of years—with each day possibly equating somewhat to the ages of geology. They believe, however, that living forms were introduced via supernatural creation throughout these various long days and that there was no large-scale evolution. Many day-age adherents are often criticized by literalists as having supposedly moved away from a literal 24-hour day reading of Genesis. Literalist critics allege that the day-agers have drifted away from the true reading because they have caved in to Darwinian pressures. However day-agers contend that Christian and Jewish scholars possessed similar views about lengthy Genesis days long before Darwin developed his theory of evolution.

The final group of progressives we include here adhere to *progressive creationism*. Progressive creationists allow for limited evolution in their thinking. They believe that the Bible allows for millions of years of evolution in which God occasionally intervened, specially creating new kinds of organisms, such as horses or humans. This interpretation thus provides a theological explanation for gaps in the fossil record—a supernaturally punctuated evolution.

One example of a leading professional progressive creationism organization is *Reasons to Believe* (RTB), headquartered in Pasadena, California. RTB produces and distributes books, booklets, magazine articles, papers, videotapes, audiotapes, and CD-ROMs. They also produce television and radio programs and maintain an Internet site. They have been planning to produce more materials targeted specifically toward children, young people, educators, and other special-interest groups.[13]

The progressive category is an interesting and important one that does not typically fit many popular characterizations of creationism. For example, the *Cambridge Dictionary of Philosophy* defines creationism as the

> acceptance of the early chapters of Genesis taken literally. Genesis claims that the universe and all of its living creatures including humans were created by God in the space of six days.[14]

Although progressives do not believe that the universe and all living creatures were created in six days, they still reject evolution as it is taught. Moreover, most of the progressive leadership would consider themselves creationists.

Literalist and progressive anti-evolutionary professional organizations combined produce a wide variety of literature written for elementary and secondary students. These materials are primarily marketed for private religious schools and for home schooling and, therefore, might appear to have no effect on other schools' students. However, the overwhelming majority of students that attend private religious schools and home schools for their elementary and sometimes junior high education, go on to attend public schools for the remainder of their K–16 years.

Literalist and progressive anti-evolutionary professional organizations also produce literature for adults. It is most likely that this literature affects students even more than the literature geared specifically to them, through a trickle-down effect. Many adults purchase and read the publications, then attempt to teach children reasons why evolution is rationally, theologically, and scientifically untenable. There is even evidence of a correlation between student attendance at occasional local anti-evolutionary meetings (i.e., seminars and evolution/creation debates) and challenges to the teaching of evolution.[15]

Theists

This group is characterized by those who contend that evolution occurred without God's (or some intelligent designer's[s']) intervention along the way to create new organisms including humans, but, for various reasons, they are against some aspects of evolution as it is taught. Persons who hold these beliefs are commonly referred to as "theistic evolutionists," "evolutionary creationists," "providential evolutionists," or "intelligent-design theorists."

There are two main divisions within the theist group. The first faction contends that the randomness involved in evolution is minimal or nonexistent. Persons in this faction basically accept evolutionary theory with the proviso that the God of the Bible, not chance, decided the human outcome by directly guiding the process. Such theists are most commonly referred to as *theistic evolutionists.*

The second faction holds that the process of evolution involves an authentic random element, but that God employed that randomness to produce a desired end—humans.[16] Adherents to this position often explain their belief by employing an analogy of a casino owner who uses chance games to produce a predictable year-end profit. In either faction, the general contention is that the dichotomy is not creation *vs.* evolution but is rather one of design *vs.* accident.

On the surface, theists may appear to have little if any cause to challenge evolutionary teaching. However, quite often in science and science education journals concerning evolution, references are made to evolutionary scientists who point out in their writings that evolution involves randomness. For example, Stephen Jay Gould and others have stated that if the history of life were to be rerun, then other organisms, not humans, would have evolved.[17] It is the teaching of this kind of randomness in evolution with which theists often take issue.

Literalist, progressive, and theist leaders have argued strongly, and many continue to argue, against the theological positions held by one another (e.g., literalists against progressives, progressives against literalists, literalists against theists, and so forth). This continues to be the case especially with literalists arguing against progressives and vice versa. In many publications, arguments have been thrown back and forth alleging that the other side is hurting the cause of Christianity with their inaccurate, unscriptural views. It is interesting to note that in the more general public arenas—secular universities—creationists rarely air their differences with one another. It is as if the various creationist camps do not want the "outside" world to know that there is significant disagreement within Christendom as to what the Bible actually states on the issue.

The theist group infrequently confronts the literalist and progressive groups. They feel that because their theistic position is so extraordinarily different from the positions of the literalists and progressives (no literal reading of Genesis whatsoever), their time is better spent arguing with evolution educators. However, intelligent design theorists occasionally point out that their views are quite different than those of theistic evolutionists. To be sure, most intelligent design advocates and many progressives say they are not creationists and avoid discussions of any pertinent theological positions they may hold. However, most people who study the anti-evolution phenomena consider intelligent design advocates as creationists, and, ironically, so do many of the leading creationists themselves. For example, the president of the leading creationist organization (ICR)

Views on Creation

Literalist	**Young universe (4,000–10,000 years), and**
	1. Literal days in biblical Genesis, and
	2. All life was supernaturally created essentially in its present form in the past 4,000–10,000 years.
Progressive	**Old universe (millions or billions of years), and**
	1. All life supernaturally created essentially in its present forms in the literal days of Genesis within the last 10,000 years, or
	2. Days in Genesis indeterminately long (millions or billions of years); living forms introduced via special creation throughout these long days; no large-scale evolution, or
	3. Days in Genesis indeterminately long; evolution did occur aided on various occasions by God's interventions of special creations of new higher order organisms.
Theistic	**Old universe; days in Genesis indeterminately long; evolution occurred; no interventions of special creations**
	1. Minimal or no randomness; evolutionary process guided by God to produce humans, or
	2. Authentic randomness; random element employed by God to produce humans.
Intelligent Design	**Some unidentified form of supernatural or extraterrestrial intelligence designed complex biological structures such as DNA and the bacterial flagellum; views were not necessarily connected to any particular religion or supernaturalism (although most often advocated by Christians)**
	1. Common descent often permissible (varies by advocate).
	2. Evolution by natural processes does not account for irreducibly complex biological structures—therefore, must have been designed.

writes that: "the trend among many Christian groups these days is to camouflage their creationism as 'Intelligent Design' or 'Progressive Creationism.'"[18] We do not agree often with the creationists, but on this point we do.

Problems with Polls

Realizing that all "creationisms" are not alike, it is easy to see how we educators can easily place students into categories (sometimes subconsciously) that do not reflect their beliefs about the subject we are attempting to teach. Likewise, it may be easy to recognize why some public opinion polls on the subject of evolution are difficult to design to take into account all types of creationist views. Polls that are ill-designed produce results that may mislead instructors in some ways.[19] The discussion that follows is of a Gallup poll and is meant to illustrate how misunderstandings concerning students' creationist beliefs can lead to false assumptions about what students find offensive or believe to be false about evolution.

One Gallup poll asked respondents to note which statement of three came closest to their views about the origin and development of man. The statements were "(1) God created human beings pretty much in their present form at one time within the last 10,000 years or so. (2) Human beings have developed over millions of years from less advanced forms of life, but God had no part in this process. (3) Human beings have developed over millions of years from less advanced forms of life, but God guided this process."[20] Such limited choices present somewhat of a dilemma for some progressive creationists. They can't choose #2 because it states that God did not have a part in the process; progressives believe God certainly did. They have problems with #3 because they believe that God supernaturally created all living things or that God at least intervened to supernaturally create when needed—He did not just guide the process. Therefore, some of these progressives choose #1 because it advocates that "God created human beings pretty much in their present form" while not believing the latter half of #1, that the creation happened "one time within the last 10,000 years or so." Therefore, if progressives who accept standard geological ages choose #1 for its creation emphasis while not agreeing with the 10,000 year age, these responses inflate the polling results for the young earth position, making it appear that there are more literalists than there may be. In other words, many progressive creationists, when asked to respond, may have chosen view #1 by default or, more aptly, considered it the least of three evils.

Also troublesome in the wording of this particular poll is the phrase "God guided this process." Although some people in the theist camp believe that God guided the process of evolution, some of them believe that His guidance is so removed from our observation (or possibly even from our understanding) that we cannot detect it. (The overwhelming majority of scientists and science instructors would find no problem with this student view when it comes to learning about evolution.[21]) However, some progressive creationists may also claim that the same terminology allows for God's guiding of evolution, which includes occasional supernatural creations of biological organisms, especially humans. This understanding of God's guidance of evolution is certainly a different understanding from that of the theistic camp. So such wording in the poll serves to blur the distinction between those who see the process as not including supernatural biological creations with those who do.

Another confounding factor in this poll (although less relevant to the day-to-day challenges of teaching evolution) is what the poll means by "God had no part in this process." On the surface, this phrase may sound atheistic. However, many theists could certainly choose this item. These theists would contend that God set up the natural laws and, given his omniscience, knew a result would be humans. However, this phrase may also be read to mean that God had no involvement in setting up the laws of nature so that things would evolve (e.g., that God is unnecessary to the whole process—evolution would have happened with or without God). Because of the item's ambiguity, both atheists and theists could have chosen this response. Many theists who believe that God set the stage for evolution consider that alone to be a creation.

A more recent (2000) national poll commissioned by the People for the American Way Foundation gives a clearer picture than the Gallup poll of what Americans may think about the relationship between God and evolution in general. In this poll, a question was asked concerning whether the scientific theory of evolution was compatible with a belief in God. The majority of Americans (68%) agreed that one does not preclude the other.[22]

Knowing that many of the issues within the evolution/creation controversy are demarcated by subtle differences may help us better understand the sensitivities students possess that hinder evolution learning.

Perceived Scientific Challenges to Evolution Teaching

Scientific Creationism

The terms "scientific creationism," "creation science," and "abrupt appearance theory" are generally used by the literalists, sometimes by the progressives, and least by theists. (When theist leaders do advocate these concepts they almost always use abrupt appearance theory.) For brevity we shall refer to all three as scientific creationism where appropriate.

Most in the scientific and education communities consider the terms "scientific creationism" or "creation science" to be oxymorons because they are as internally inconsistent as "scientific religiosity" or "religion science." However, literalists see these terms as distinguishing scientific creationism from what they perceive to be "scientific evolutionism." Naturally, the latter term is not recognized in the scientific community because evolution is, in fact, scientific. Therefore, "scientific evolutionism" is as redundant as "scientific science." Nevertheless, literalists see their terms as distinguishing between what they perceive as their good science vs. the scientific establishment's bad science (which is virtually anything that supports evolutionary theory).

Perhaps literalists also use this terminology to appeal to educated persons. For adults, most polls indicate that the higher the education level attained in general, and the higher the level attained of science education in particular, the higher the understanding/acceptance of evolution (see Charts A–C).

Additionally, literalists often use this terminology as an attempt to distinguish between what they perceive to be the scientific evidence of creationism and what they believe based on the Bible. It is a literalist contention that regardless of how one interprets the Bible or which religion one believes (with a few possible exceptions), the scientific evidence alone will compel any believer who examines the evidence with an open mind to conclude that organisms did not evolve.

Progressive creationists also espouse a form of scientific creationism but do not usually call it such because this would equate them with the

CHART A

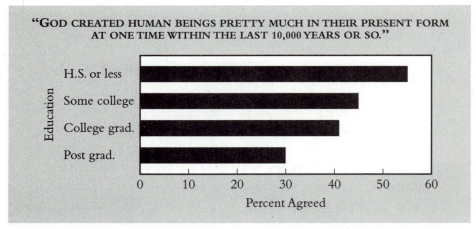

"GOD CREATED HUMAN BEINGS PRETTY MUCH IN THEIR PRESENT FORM AT ONE TIME WITHIN THE LAST 10,000 YEARS OR SO."

SOURCE: The Gallup Organization, Inc. Survey dates: August 24–26, 1999

CHART B

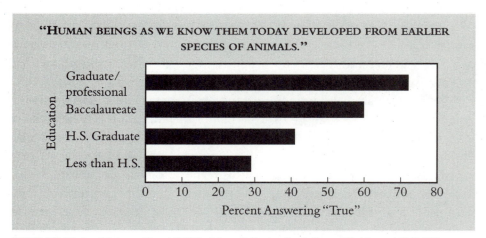

"HUMAN BEINGS AS WE KNOW THEM TODAY DEVELOPED FROM EARLIER SPECIES OF ANIMALS."

SOURCE: National Science Board (1996). *Science & Engineering Indicators.* Washington, DC: U.S. Government Printing Office (NSB 96–21)

CHART C

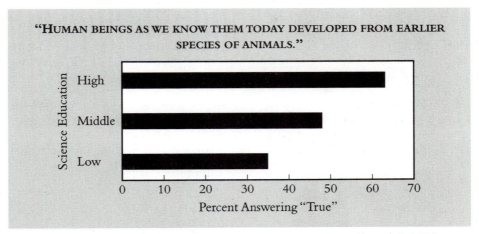

"HUMAN BEINGS AS WE KNOW THEM TODAY DEVELOPED FROM EARLIER SPECIES OF ANIMALS."

SOURCE: National Science Board (1996). *Science & Engineering Indicators*. Washington, DC: U.S. Government Printing Office (NSB 96-21)

literalists. However, like the literalists, the progressives contend that apart from relying solely on biblical rationales, they have sufficient scientific arguments and evidence to support the beliefs as discussed previously.

The "scientific" (and often biblical) positions of the professional progressives and literalists are often at great odds with one another. The presidents of the largest young-earth organization and largest progressive creation organization, the Morrises of the Institute for Creation Research, and Hugh Ross of Reasons to Believe, respectively, have argued strongly back and forth in print for years. Amazingly (at least to those of us who contend that evolution is scientifically accurate), they seem to be offended most by each other's contention that the other's positions allow for some evolution to have taken place. Ross writes:

> Such creationists [referring to Morris] brand day-age proponents, like myself, who deny any significant biological evolution over time scales long or short, as evolutionists, while they themselves seem to concede substantial biological evolution over very short time scales.[23]

Literalists certainly do not like others painting a picture that they concede any biological evolution over any amount of time, let alone substantial evolution over short periods of time. Likewise, progressive creationists certainly do not like being labeled, of all things, evolutionists.

Students and others who approach science instructors with the plan of interjecting creationism into a course generally attempt to do so through the arguments of scientific creationism and rarely through direct biblical rationales. The professional creationists contend that there is a demarcation between scientific creationism and what they term "biblical creationism." The Institute for Creation Research's educational philosophy states that

> A clear distinction is drawn between *scientific creationism* and *biblical creationism* but it is the position of the Institute that the two are compatible and that all genuine facts of science support the Bible. ICR maintains that scientific creationism should be taught along with the scientific aspects of evolutionism in tax-supported institutions.[24]

Although people attempting to introduce creationism into science courses generally do so using a "scientific" approach, their underlying motivation is almost always religious. Many in the science and science education communities suspect that the underlying motivations of professional creationists, however, are eventually to interject religious doctrine into public school science classrooms rather than just to counter "bad" science and teach "good" science. These allegations are partially based on the fact that most of the leading creation organizations in the United States are literalist fundamentalist Christian institutions. Such institutions are generally not known for their biblical mandate to teach good science but rather to spread the gospel. After all, whether leading creationists admit it or not, a 10,000-year-old earth with immutable "kinds" of organisms is linked to a theological dogma. However, the surface arguments creationists put forward may appear to sound somewhat nonreligious or even scientific in nature. This does not mean that these students are intentionally disguising their religious motivations. They just think that they need to teach their science instructors good science so that they will no longer teach other students scientific inaccuracies, which are also biblical inaccuracies. Of course, if a science instructor eventually concluded that today's basic life forms were spontaneously created, most creationists would feel very happy but also biblically obligated to share ideas with the instructor about the Supreme Being who was responsible for the spontaneous creation.

If evolution is not accurate, then the only other alternative is supernatural creation. Is this statement true? This dichotomy is often debated, but its advocates are not just the creationists. For example, MIT's Steven Pinker, certainly no creationist, states his views on the matter quite succinctly:

> And here is the key point. Natural selection is not just a scientifically respectable alternative to divine creation. It is the *only* alternative that can explain the evolution of a complex organ like the eye. The reason that the choice is so stark—God or natural selection—is that structures that can do what the eye does are extremely low-probability arrangements of matter.[25]

Pinker's statement helps illustrate that creationists are not the only ones popularizing a position of only two respectable choices: creation or evolution. (There are alternatives to this dichotomy, as we shall soon see.)

Because many people believe that life arose on earth by either creation or evolution, creationists present arguments against evolution in the hopes of demonstrating to others that it therefore must have arisen by creation. In this context, creationist arguments include:

1. biological life could not have developed from the inanimate via natural processes,

2. the diversity of life we see today could not have evolved from lower forms of life,

3. no evolution can occur beyond, roughly, the phylogenic level of "family,"

4. humans did not evolve from lower animals and, since their creation, have always possessed all the basic human characteristics of today's human beings ("self-image, moral consciousness, abstract reasoning, language, will, religious nature, etc."[26]),

5. the earth and the universe are not billions of years old but rather 10,000 years old or less,

6. most sedimentary rocks containing fossils are the result of a global flood occurring less than 10,000 years ago, and

7. all organisms when they were originally supernaturally created were created perfectly and over time have experienced physical degeneration.

While most creationists and some noncreationists contend that there are only two positions—evolution or the traditional creationisms—others disagree. Philosopher of science Robert Pennock argues that merely disproving evolution does not automatically usher in the traditional creationisms as the only alternative. His argument shows that creationists must do *more* than disconfirm evolution in order to have their position accepted. If creationists want equal time in the science classroom, alternative explanations beyond theirs would need equal time as well. As an example, Pennock uses the Raëlian movement. Followers of this movement believe in a secular, naturalistic origin that is neither evolution nor creation. (The Raëlian movement claims over 50,000 members in 85 countries.[27]) They basically believe that a nonsupernatural extraterrestrial intelligent designer has run a long-term experiment to create life on earth. (This second alternative to evolution also fails as science, of course, and therefore has no place in a science curriculum.[28])

It is a fairly safe assumption that while Christian creationists are not satisfied with evolution being the confirmed scientific theory, they also would not be happy with Raëlian beliefs being presented in the science classroom. So, ironically, the "fairness" of equal time in science classrooms for nonevolutionary explanations that the creationists so passionately seek might result in teaching even more explanations that literalists and progressives find offensive. Pennock also demonstrates that the Raëlian faith is similar to the faith of a more recently organized type of creationists, who prefer to be called intelligent design advocates.

Intelligent Design

In the past decade, many groups advocating different forms of creationism (than we have previously discussed) have become popular. These groups argue that science and science education should include concepts that they call, among other things, "intelligent design theory," "initial complexity theory," and "theistic science." Similar to groups that advocate some form of scientific creationism or progressive creationism, the intelligent design (ID) groups contend that there are compelling scientific arguments and evidence that would lead rational thinking people to conclude that evolution did not occur in the manner normally taught in science classes. However, unlike the typical proponents of the creationisms discussed previously, most ID proponents strongly distance themselves from the label "creationist" and often from any overt connections to religious motivations or rationales for their ID conclusions.

Nevertheless, most evolution educators who have studied the arguments and activities of the ID leaders have concluded that their actions are simply attempts to create a new delivery system for creationism. They also conclude that ID leaders want their ideas, particularly supernatural hypotheses, special (individual) revelations, or "theistic science" to be considered by the scientific community and ultimately injected in public school science curricula.

This new strategy is likely being used by ID advocates for two primary reasons. First, past attempts to introduce overtly biblical doctrine into public school science curricula have failed due to promotion of one particular religious viewpoint—a literalist interpretation of Genesis. Second, "science" overtly labeled "creationism" has been thoroughly discredited in the scientific community. Therefore, as a delivery system for theistic science, creation science has little chance of inclusion in public school science curricula. So by distancing ID from labeling that has failed in the past, the ID advocates might feel as though they have a better chance of accomplishing their goals. However, no matter what the "packaging" of ID arguments, creationists still remain similar at their core: Evolution, as taught in science classrooms, did not occur. Therefore, they contend we should include supernatural cause (or intelligent extraterrestrial causation) as an explanation in science.

To some extent, the strategy of intelligent design groups has already been successful. Financial support of their endeavors appears to be growing. Large numbers of people have attended talks by ID leaders, who sometimes draw audiences of up to nearly one thousand. Copies of Phillip Johnson's book *Darwin on Trial,* an anti-evolution book that has been praised by leaders of the ID movement, have been sent to biology teachers throughout the state of Alabama, courtesy of the state's governor. And for the first time in recent history, a major trade book publisher (The Free Press, A Division of Simon & Schuster Inc.) has published overt intelligent design material. The book is Michael Behe's *Darwin's Black Box.*

In summary, all creationisms are not alike. Creationists' ideas vary significantly with regard to both their religious and nonreligious rationales for rejecting evolution. Likewise, creationists' ideas vary considerably in the extent to which they reject evolution. A better knowledge of these views may provide science instructors with a deeper understanding of their students' rejection of the fundamental concept in the life sciences, which in turn should allow for improved pedagogy.

The General Public

It seems that much of the general public, including many secondary and undergraduate science students, unfortunately do not have even a rudimentary understanding of how the scientific community operates, the purpose of scientific journals, how the peer review process works, how scientists compete for research funds, and so forth. Therefore, when North Americans hear or read about differing scientific views concerning the specifics of evolutionary mechanisms, many misunderstand—and instead believe—that the fact of evolution has not been firmly established. National polling indicates that almost half of Americans (49%) think that evolution is "far from being proven," while, by comparison, less than 10% think the same way about Einstein's theory of relativity.[29] Naturally, these discussions in the scientific community about evolutionary mechanisms have nothing to do with creationists' arguments, nor do they involve the creationists themselves. But such misunderstandings most likely help to engender much of the rejection of evolution that is reported in various polls. To compound the matter, it is unfortunately the case that most Americans probably just don't think much, if at all, about evolution. And some who spend limited time pondering it simply suspend judgment.

Even many university freshmen think that scientists question evolution's occurrence if they hear or read of differing scientific views concerning the specifics of evolutionary mechanisms. Although most of the major scientific organizations, including the prestigious National Academy of Sciences, have published official statements endorsing the scientific factuality of evolution, often students only need encounter a creationist or two with the "proper" scientific degrees and profession to decide that the factuality of evolution is unsettled in science. It appears for them, as for many Americans, that a less than 100% agreement among all science Ph.D.'s about evolution's scientific factuality is reason enough to tell stories to others that evolution has yet to be established.

Often public creation/evolution debates at universities and the occasional televised one assist in fueling the general public's perception that a scientific controversy over evolution rages on. For many who listen to such debates, the take-home message does not involve the specific scientific arguments of the debate but rather that both sides had people with Ph.D.'s in the sciences who sounded rather rational. Unfortunately, such debates typically engender the perception that if the occurrence of evolution were a settled issue in science, then scientists would not be debating it. (We have witnessed many formal creation/evolution debates in which it was our perception that the creationists clearly lost the contest. Yet in speaking with members of the audience after the debate, we learned that our perception was not necessarily shared.) It seems that the combination of little if any scientific training, a misunderstanding of the way the scientific community operates, and the possession of a belief that evolution just seems intuitively impossible, do much to inhibit the possibility of one concluding that the evolutionary explanation is the only, or at least the superior, scientific explanation.

The situation for the fundamentalist Christian lay-public is often similar. If they are told something about science from someone who has a Ph.D. in science, then they believe that person must be qualified to speak on behalf of science. This is especially the case when a scientist's non-science views are in line with what the person already knows to be true (e.g., that the Bible is the word of God). The creationist is generally introduced, whether in person or in text, as having received his Ph.D. in science from a known university, currently being a college professor, and being published in various recognized scientific journals (although rarely, if ever, is it mentioned that those publications have nothing to do with the creation/evolution controversy). It is therefore assumed by fundamentalist Christian lay persons that the creationist is someone they can trust to tell them where "some members" of the scientific community are incorrect about their science (i.e., evolutionary science). Therefore they believe that they have learned the difference between good science (creation) and bad science (evolution).

Category Solution: Nature of Science

One of the most popular methods espoused by mainline science educators to help alleviate creationist-perceived tension between biblical-creation accounts and evolution (or religion and science in general) is to teach that science is different from other ways of knowing. This method is used by many in an attempt to have creationist students understand that there are different rules for different games; that when we play the game of baseball we don't incorporate rules from ice hockey. It doesn't necessarily mean that baseball is a superior game to ice hockey; it is just different.

Likewise, science instructors usually tell students that science has different rules, methods, and epistemology(ies) than say, religion, art, history, or personal experience. For example, one rule (philosophical) modern science uses is often referred to as *methodological naturalism,* whereby scientists seek naturalistic causation for nature's phenomena. In other words, scientists could examine whether gravity has an effect on the ocean's tides but could not make a ruling on whether God has an effect on the tides. Science looks for just naturalistic causes for what caused X, while some creationist religions may consider supernatural causes for what caused X. Additionally, since one way of knowing cannot judge other ways of knowing, no one way of knowing can be used to judge the accuracy of the others.

Often such seemingly simple, effective, and logical methods used to separate science from religion are considered unreasonable by creationist students—or at least confusing by many. Instructors are often baffled that creationist students can't seem to, or are unwilling to, grasp this solution in order for creationists and instructors of evolution to live in harmony in the classroom. The problem is that these students believe that they would be able to rule which "game" is better—baseball or ice hockey (evolution or creation)—not by looking through the rules of one of the games, but by biblical rules. If they perceived that a passage in the Bible communicated the superiority of one game over the other, then the analogy of comparing games to ways of knowing is not appropriate or useful in their minds.

Figure A is an analogical representation of viewing four games independently, one from another. Figure B shows that one game's rules cannot be viewed through another game's rules. However, many creationist students actually do view the games of baseball, ice hockey, basketball, and football through the lens of their interpretation of the Bible (usually including their churches' traditions and practices). Even some graduate students we know at leading research institutions consider games using cards, pool tables, and Ouija boards (but not necessarily involving gambling), to be clearly inferior to baseball, ice hockey, basketball, and football. These judgments are arrived at through a lens that does not approve of the games, which is illustrated in Figure C.

However, many would rightfully point out that the game analogy in Figure A did not include religious rules, only rules concerning the different games of baseball, ice hockey, basketball, and football. After all, the only reason the students were able to make a ruling on those games is that they brought in religious rules outside of the analogy. This is precisely the surprising element for many science instructors: For many creationist students, all analogies will be subjected to their biblical rules.

Other science educators do not use analogies to distinguish between science and other disciplines but simply use direct instruction, explaining that art, history, personal experience, religion, and science are different ways of knowing. The lesson is presented in a manner that treats all areas

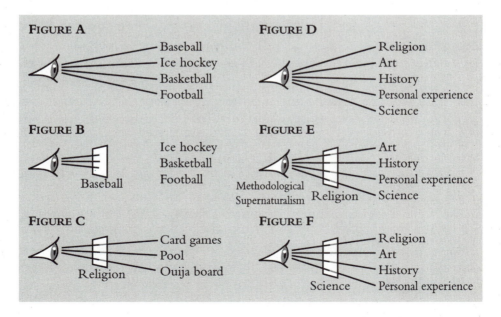

with equal respect, attempting to show that while science may currently conclude X about a question in the science domain, and religion may conclude Y about a question in the religion domain, one cannot judge the conclusions of the other because they operate in different domains with different questions (Figure D). Many science educators consider this presentation relatively fair and reasonable.

Such a view is an anathema to many creationists. They view the areas of art, history, personal experience, and even science, through a biblical lens (Figure E). This biblical lens is modified by a synergism of such factors as Christian beliefs engendered by Christian literature, church leaders, local church doctrine, larger denominational doctrine, personal interpretation of Scripture, and parental biblical teachings. Confusion in the classroom results when the instructor contends that Figure D is an acceptable position, while the creationist student believes that Figure E is correct. The attempt by science instructors to remove the religious lens from directly in front of the eye and move it to a position next to science, art, history, and personal experience is just not "doable" in many cases. Some argue that the promotion of Figure D in the classroom is often unsuccessful because it is basically an attempt to go against the student's worldview, which has been instilled by his or her cultural upbringing over many years. However, Figure D may be useful (if appropriate in the setting) to teach students how some of religious faith represent their resolution of religion and science. Nevertheless, just because it does not resolve some students' perceived evolution/creation conflict, it can show them how others resolve the matter. Such a presentation may be appropriate in the science classroom as part of philosophy of science in relation to knowledge in general. (Naturally, additional science/religion resolutions might also be offered to the students.)

The philosophical underpinnings of science are important for students to understand, and methodological naturalism (MN) is one of the fundamental rules governing modern science. Although most leading creationists conduct the overwhelming majority of their science without reference to supernatural causation (see Chapter 7)—therefore applying methodological naturalism to their work—most of these creationists would argue that methodological naturalism should not be a necessary element in doing modern science. But what about all those Christian professors who contend that MN *is* appropriate in science and therefore have no trouble accepting evolution as the scientific explanation for the diversity of life on earth? What do the anti-evolutionists think of them? The leading anti-evolutionists almost universally state that these professors are deceiving themselves, because they have been convinced of the "incorrect" philosophical underpinnings of science.

In fact, leading ID anti-evolutionist Phillip Johnson feels that "When MN is understood profoundly, theism becomes intellectually untenable."[30] One would expect that those scientists who are devoutly religious and understand "profoundly" the pragmatics of MN would have some significant disagreements with Johnson—and they do. (We will not take the space in this book to air the significant disagreements those scientists in the Christian community have with Johnson's contentions. We will only say that the nay-sayers are quite numerous.)

In any case, science instructors' attempts to lessen the resistance to evolution by the categorical means in Figure D and by incorporating MN have been extensively criticized by leading anti-evolutionists. They contend that this approach does not simply teach our students MN but also teaches metaphysical naturalism (MPN). MPN is defined in the following quotation. We have chosen the definition of a prominent anti-evolutionist because this is most likely the definition on which many creationist students and parents base their opinions.

> According to naturalism, what is ultimately real is nature, which consists of the fundamental particles that make up what we call matter and energy, together with the natural laws that govern how those particles behave. Nature itself is ultimately all there is, at least as far as we are concerned. To put it another way, nature is a permanently closed system of material causes and effects that can never be influenced by anything outside of itself—by God, for example. To speak of something as "supernatural" is therefore to imply that it is imaginary, and belief in powerful imaginary entities is known as superstition.[31]

Figure F represents what Johnson fears: the examination of religion, art, history, and personal experience all through the lens of science—MN science. Here we have the case of not merely being accused of teaching students MN but actually leading them to MPN; that is, moving students from Figure D to Figure F. It's somewhat of a slippery slope argument: Because science has been so successful, if students learn that MN philosophically underpins modern science, then students will ultimately view all of life through the lens of MPN. The great fear of many leading creationists and the misunderstanding of many creationist students is that accepting the MN rules for doing science will necessarily lead to a supernatural-less/Godless worldview.

Creationists attempt to make their case by claiming that much of their fear is engendered by scientists who are well known by the general public and who intermix MPN with MN when explaining how science operates in the conclusions it draws. Such luminaries as Crick, Dawkins, Gould, Hawking, Sagan, Leakey (Richard), and Suzuki have been accused of furthering the idea that because MN works so well in science, it would naturally follow that MPN works well in all of life. Naturally, the anti-evolutionists contend these scientists don't always explicitly state these views, but nevertheless believe the MPN message is there, clearly and implicitly.[32]

So what about science classroom MN sliding into the feared MPN? The large number of devoutly religious scientists who have no trouble underpinning their science activities with MN is overwhelming empirical evidence that the slippery slope is really not that slippery at all. Scientists can do naturalistic science while being devoutly religious. In any case, the science classroom should teach how science is done, and MN underpins modern science.

An understanding of the creationist culture from which students may come should help science instructors with the teaching of evolution. Such understanding should yield an increased knowledge of the origin of many science misconceptions students hold, explain why some teaching strategies won't work and others may, improve communication about evolution, and even allow instructors to better teach evolution in a manner less threatening to students.

3

Why Students Reject Evolution: Religious Reasons

"I see no good reason why the views given in this volume should shock the religious feelings of any one."

—Charles Darwin, 1872, *The Origin of Species*[1]

Alas, if only it were true for everyone. However, that has never been the case. A considerable number of people do feel shocked by the theory of evolution and believe they have good reason to be appalled. And science instructors are the ones most often teaching those offending views.

In this chapter we will present religious rationales for why some students, students' parents, and possibly fellow instructors or administrators may find evolution to be theologically unacceptable to them. The Bible is the one common factor among almost all Christian creationists, and Christian creationists constitute the majority of those who openly challenge the teaching and accuracy of evolution as taught in science courses.[i] Discussing religious rationales can be a difficult task because of the tremendous variation of theological views even within the narrow traditions of creationist Christianity. One easily runs the risk of unintentionally misrepresenting some factions. However, lessening this risk greatly is the

---

[i] Eighty percent of Americans believe the Bible is the literal or inspired word of God (Gallup News Service 2000).

fact that the majority of creationists who most often overtly challenge the accuracy of the evolution being taught, fall within a certain theological tradition. In this chapter, we focus on this group—the literalists.[2]

Although literalists generally are considered biblical fundamentalists, and progressives generally are considered more moderate, it should not be assumed that progressives are any less confrontational than literalists against the evolution normally presented in science textbooks. Progressives can be every bit as passionate as literalists when it comes to proclaiming why the evolution considered factual in the scientific community is extremely inaccurate. They share with literalists many basic beliefs about the Bible, but, as outlined previously, have some relevant, significant differences when it comes to learning about evolution. However, understanding the religious rationales for why the literalists reject evolution can aid in under-standing many of the reasons why progressives reject evolution. Literalists have numerous religious rationales for rejection, while progressives share fewer of those religious rationales.

Theists do not have as many religious objections to evolution being taught as either the literalists or progressives do because they read the bib-lical creation account as mostly, or entirely, metaphorical. Their religious objections generally concern the extent to which randomness is involved in evolution, as discussed in Chapter 2. We will not be exploring the the-ology and exegesis (interpretation of the Scripture) involved in theists' objections to some parts of evolution education. In short, theists believe that the Bible indicates that God created everything, but that evolution is permissible (at least most of it). As a consequence of these two linked beliefs, many theists want science to include theistic causation.

Whether literalist, progressive, or theist creationist, all contend over-whelmingly that the Bible is a very special communication to mankind from God that has bearing on the teaching of evolution.

Students' Biblical Scholarship

The vast majority of Christian creationists encountered by science instruc-tors in the classroom certainly are not biblical scholars, and they generally would be the first to admit this for two reasons: It is usually true and humility is a biblical virtue. They claim to believe in the Bible most of the time, but they know very little about its contents—usually to the great disappointment of their clergy. However, while creationist students may not be able to cite multiple biblical chapters and verses to substantiate their point of view on a theological argument, many of them understand the basic principles or concepts of the particular theological tradition to which they belong. Likewise, many students don't understand the precise biblical reasons for their rejecting evolution, but nonetheless understand that it is against their church teachings (in those cases where their churches, in fact, do reject evolution). Often the students don't know exactly where they heard the biblical justification for denying evolution—maybe from church leaders, local church teachers, parents, or friends. Whatever their sources are, we have found them often to be tertiary. Their ideas can sometimes be traced back to more specialized sources—the creationists' publications themselves—which serve as secondary sources. Those sources, in turn, explicitly point to their alleged primary source, the Bible.

Ironically, evolution instruction motivates many students to study (or to increase their study of) the Bible and to purchase creationist products. Why? Quite often students have creationist beliefs even before they understand the underpinnings of such beliefs—they simply regard their beliefs as intuitive. However, when these beliefs are challenged, students are often motivated to seek greater biblical justification for why they believe what they believe—for the purposes of defending their beliefs.

Biblical Rationales

Many noncreationists, and to some extent many creationists, think that any apparent biblical problems with evolution are confined to just the few verses about creation in the first chapter of Genesis. However, most creationist church leaders and leading creationists are quick to claim that references to a literal Genesis creation are not relegated to a simple reading of just the first chapter of Genesis. The following are some of the rationales that *creationists give* for their believing that the Bible does not allow for evolution.

"In six days the Lord made heaven and earth, the sea, and all that in them is, and rested the seventh day: wherefore for the Lord blessed the sabbath day, and hallowed it" (Exodus 20:11).[3] This is corroborating evidence for the position that the authors of the Bible mean to communicate to mankind that everything we see was created in six literal days as detailed in Genesis. Because this seven-day week pattern is adhered to for modern periods of work, and, more importantly, because the sabbath day currently observed is 24 hours long, this passage clearly indicates solar days and not possible epochs. Therefore, sudden creation occurred, not evolution.

"As by one man sin entered into the world, and death by sin" (Romans 5:12). The one man being referred to is Adam and it is by his original sin in the Garden of Eden that death began. Death (at least animal death—including humans) did not exist on the planet prior to Eve's encounter with the serpent and Adam's subsequent sin. Because there was no death, there could be no survival of the fittest, no natural selection, no extinction, and therefore no evolution.

"For God sent not his Son [Jesus] into the world to condemn the world; but that the world through him might be saved" (John 3:17). Saved from what? God sent Jesus into the world to save mankind from the repercussions of "the fall"—the fall from a sinless existence into a sinful existence that caused harsh spiritual judgment. The fall occurred because of Adam's sin, and physical death and decay for the planet were two of the results. If death had occurred prior to Adam, as evolution requires, then there would be no need for salvation via Jesus.

"Thus the heavens and the earth were finished, and all the host of them. And on the seventh day God ended his work which he had made" (Genesis 2:1,2). God finished his work on creation long ago; he is no longer creating any new kinds of animals. Therefore, if God did use a type of evolution in the past to create different kinds of animals, it could not be operating today. So since the evolution being taught in science classrooms claims that evolution is still operating, it can't be accurate, because God is no longer creating new kinds of animals.

"And God saw everything that he had made, and, behold, it was very good" (Genesis 1:31). God's declaration that the creation was "very good" indicates that at the time of his creative works the earthly organisms experienced no suffering. This means that the suffering (and death) involved in the survival of the fittest must not have been occurring, and therefore, no evolutionary processes could have been involved in the origination of earth's organisms.

"And the Lord God caused a deep sleep to fall upon Adam, and he slept; and he took one of his ribs, and closed up the flesh instead thereof. And the rib, which the Lord God had taken from man, made he a woman, and brought her unto the man" (Genesis 2:21,22). The first woman was created from Adam's rib while he slept—not over millions of years of evolution.

Sometimes students bring up the interpretation of Scripture to dispute scientific facts that most of us take for granted, such as the extinction of the dinosaurs greatly preceding the arrival of humans. Most creationists believe that humans and dinosaurs coexisted from the beginning. Creationists do not always rely exclusively on their reading of Genesis for such beliefs but also rely on what they think the Bible was referring to with its words "dragon" and "behemoth." The Bible describes a "behemoth" that "moveth his tail like a cedar" (Job 40:15–24) and discusses other dragon encounters elsewhere (e.g., Isaiah 34:13, Micah 1:8, Malachi 1:3). Many creationists believe these biblical animals simply possess antiquated names for organisms that we today refer to as dinosaurs. Since they think that no other organisms move their tails like cedars, this is clear evidence to them that humans and dinosaurs coexisted. Creationists often buttress this argument with nonbiblical "historical" accounts such as those of ancient people who had encounters with dragons/dinosaurs.[4]

The idea that there cannot be too much change allowed over time for any particular "kind" of organism is another example of a creationist argument against evolution directly derived from the interpretation of particular words in the Bible. The word "kind" is used multiple times in Genesis, for example,

> And God said, let the earth bring forth the living creature after his kind, cattle, and creeping thing, and beast of the earth after his kind: and it was so. And God made the beast of the earth after his kind, and cattle after their kind, and every thing that creepeth upon the earth after his kind. (Genesis 1:24, 25)

Many creationists read this passage as a form of reproductive integrity that God encoded in organisms from the beginning, which inhibits organisms from evolving "significantly." The creationists argue among themselves as to how significant this change within kind can be, but most literalists and progressives don't allow the change necessary for an evolutionary explanation for the diversity of life. (We will further discuss human/dinosaur coexistence, what creationists mean by "kind," and how much evolution is allowed in the next chapter.)

There are other verses that creationists feel support their anti-evolution convictions. The previous verses and explanations are but a few examples. Naturally, there are also many who believe in the Bible but interpret those same verses as saying nothing against evolution.

A small number of college evolution instructors have gone so far as to encourage creationist students to better understand the Bible, in the hopes that these students will realize that evolution is biblically permissible. For example, they try to point out to the students that the creation account in the first chapter of Genesis is contradictory to the creation account in the second chapter (Genesis 2:4–19). In the eyes of many believers, the second chapter clearly tells a different story of creation from the one starting in the first verses of Genesis. For example, Adam was created first in Genesis 2 (before animals) and only at the end of Genesis 1 (after animals). In other words, they believe that because there are two *different* creation accounts back to back in Genesis, this is evidence that the two stories should be read metaphorically, not literally.

However, as we've mentioned previously, although creationist students may not have biblical justifications at their fingertips, they typically get them from creationist authorities when they think it is necessary. They return with an explanation similar to the following: There really is no contradiction. The first chapter gives a summary of the creation week, while the second chapter provides some of the specifics. Furthermore, the second chapter's account most likely had different authorship (possibly Adam himself) and thus presents a different perspective. Combining this explanation with some Hebrew/English retranslation of the existing English text in the second chapter makes the contradiction conveniently disappear in their minds.

Such creationist hermeneutical acrobatics are common in spite of the many exegetical problems these acrobatics create, such as reading the first chapter of Genesis employing strict literalism and then employing far less literalism in reading the second chapter. Noncreationist biblical scholars have pointed out such problems to creationists. Nevertheless, the professional creationist community has a multitude of deeply held biblical justifications that appear reasonable to them.

Engaging creationists in a conversation about the biblical underpinnings for their beliefs most likely will not bring about resolution to the issue but will, instead, escalate the discourse. Even if it did result in resolution, such action initiated by science instructors probably would draw the attention of some creationists (and noncreationists) both inside and outside of academia. Creationists would consider it highly inappropriate that science instructors advocate ways to interpret religious documents. Interestingly, many creationists that would vehemently protest against science instructors teaching biblical exegesis (that differs from theirs) are the same creationists who want their religious ideas taught by science instructors (i.e., creationism).

It comes as no surprise to most people that there are many ways to interpret the Bible. Creationists' interpretations are, of course, contained within their publications. In fact, it is usually obvious which books are creationist and which ones are not by their titles. However, there are some extremely popular books that are not recognized by most people as possibly being creationist—study Bibles.

Study Bibles

Creationist biblical justifications are often engendered by reading Bibles— but the justifications do not necessarily come from Scripture itself. Popular among students in American Protestant Christian culture, "study Bibles" contain the entire text of the Bible but also have added commentary, which usually is written by well-known Protestant Christian leaders or theologians. Skimming through a study Bible, one generally sees approximately 2,000 pages of what appears to be a typical-looking Bible with its book, chapter, and verse notations. However, annotator notes are at the bottom of almost every page. These notes are attempts to help the reader better understand the biblical verses printed above and to direct the reader to other related passages in the Bible. In addition, the annotator may provide historical or modern examples to which he thinks (rarely, *she* thinks) the biblical verses may be referring, or situations to which counsel given in a verse may be applied. These notes can range anywhere from a sentence or two explaining nearly a page of biblical text, to an entire page or more of notes illuminating just 5 biblical verses (or approximately 10 sentences).

The *theological* position of most of the Protestant Christian community is that the biblical text itself is the utmost authority and that any commentary on the Bible is only a human attempt to understand the word of God. Therefore, the commentary cannot be considered authoritative. However, much of the lay Christian community has significant respect for and confidence in those who authored the commentary in their study Bibles. This attitude—that the commentary is of great importance—often is not just a product of congregation members' thinking but is sometimes engendered by the annotators themselves. For example, one study Bible that has been popular for many years is *The Ryrie Study Bible*, which was annotated by Charles Ryrie, Professor Emeritus of Systematic Theology at the Dallas Theological Seminary. This institution is well known and influential among evangelical Christianity. At the beginning of his study Bible are the following instructions:

> Every time you read this Bible, whether carefully or casually, be sure to look at the notes at the bottom of the page. These are designed to illuminate and help you understand the verses you are reading.[5]

Which study Bible a person carries to church or uses at home is often a sign to others of his or her theological stance, which of course is quite important to many believers. If a person attends a very conservative church, as many creationists do, and is seen carrying a liberal study Bible at church, others in the congregation might think that person is somehow missing the mark—is not reading the "good" material. Some of the more caring (or more forward) members in such a church will often politely but firmly recommend a study Bible that they and the church believe to be the better text, possibly even pointing out many of the shortcomings inherent in the study Bible that person is currently using. If the person continues to use the more liberal study Bible as a resource, it would not be surprising to have a fellow member of the church present the "better" study Bible as a gift.

Although many study Bibles may seem to be scientifically benign, they can actually act, in part, as "stealth" delivery systems for anti-evolution dogma. Even when the vast majority of people think a particular biblical verse could have no possible bearing or relevance to evolution, the creationist study Bible commentators often attempt to make a link. Henry

Morris has authored a study Bible containing over 1,600 pages dedicated to "defending the faith from a literal creationist viewpoint" (*The Defender's Study Bible*).[6] This study Bible may be the most important publication in recent history that attempts to tie the entire Bible, verse by verse whenever conceivably possible, to literal creationism.

There are many other popular creationist books that are wholly or partly dedicated to discussing various sections of the Bible in an attempt to assemble scriptural support and relevance to literalist creationist theology. However, no books of which we are aware systematically cover the entire Bible with such a focused intent as Morris's study Bible. For example, the book of Matthew contains an admonition concerning how to detect false prophets: "A good tree cannot bring forth evil fruit, neither can a corrupt tree bring forth good fruit" (7:18). Typical Protestant Christian commentaries concerning this verse generally state that false prophets cannot produce good results; by the same token neither could a true prophet produce bad results. In juxtaposition, consider Morris's comments on the verse:

> This criterion of fruit inspection can be applied both to individuals and to systems. For example, the evolutionary philosophy has produced no good fruits whatever. Instead, it has generated atheism, humanism, communism, fascism, racism and all manner of evil fruits. It, therefore, is a false and evil philosophy.[7]

This commentary might be shocking to many science educators, but this example from Morris is useful in illustrating the extent to which some prominent creationists will go in their attempts to relate biblical verses, which seem unrelated, to an anti-evolution position. His language is blunt and deliberate; this is characteristic of the rest of his study Bible.

The most important point about this discussion on study Bibles is that many Christians who normally would never purchase creationist literature *are* likely to purchase a study Bible. Unlike the Morris study Bible, the titles of many other study Bibles whose annotators infuse creationist notes might not indicate to the purchaser that the Bible's commentary presents a creationist viewpoint. And the purchaser may very possibly be buying the study Bible for another person.

There is a long tradition in conservative Protestant Christianity of giving Bibles as gifts, especially to young people—Christians and non-Christians alike. This custom is so common that many Bibles and study Bibles have a page specially printed at the front on which presenters of the gift Bible can fill in their name, date or occasion, and to whom the Bible is being presented. Those who knowingly receive a creationist study Bible are more likely to read it than other creationist publications because most other creationist publications have few uses for the reader other than to learn about creationism, how to defend it, and what is "evil" about evolution. (Creationist leaders would contend that these other publications *are* about other Christian concerns because, they assert, literal creation is foundational to Christianity and pervasive throughout the Bible.)

Whether read at church or elsewhere, creationist study Bibles are significant learning tools in comparison to other creationist publications because of the medium in which the creationist message is delivered. It should be noted that many creationists would likely find such a contention to be ironic, for they would contend it is not their commentary in study Bibles that delivers the creationist viewpoint but rather that it is the biblical text itself. However, creationists, noncreationists, Christians, and non-Christians alike would probably agree that when young people encounter the word "day" in the 5th verse of Genesis, "And God called the light Day, and the darkness he called Night. And the evening and the morning were the first day," there is a greater likelihood of their coming away with a literalist creationist interpretation of this passage after reading study Bible commentaries like the following:

> Since daytime closes at evening and the night ends with the morning, the phrase indicates that the first day and night had been completed. Evening and morning cannot be construed to mean an age, but only a day; everywhere in the Pentateuch [the first five books of the Bible] the word *day*, when used (as here) with a numerical adjective, means a solar day (now calibrated as 24 hours). *The Ryrie Study Bible*[8]

> The use of "Day" (Hebrew *yom*) in Genesis 1:5 is its first occurrence in Scripture, and here it is specifically defined by God as "the light" in the cyclical succession of light and darkness which has, ever since, consti-

tuted a solar day. Since the same word is used in defining all latter "*yoms*" as used for this "first" *yom*, it is incontrovertible that God intends us to know that the days of creation week were of the same duration as any natural solar day. *The Defenders Study Bible*[9]

This cannot mean an age, but only a day, reckoned by the Jews from sunset to sunset. . . . "Day" with numerical adjectives in Hebrew always refers to a 24 hour period. *The MacArthur Study Bible*[10]

When colleagues have been shown such quotations as above without their references, they naturally thought they were taken from overtly creationist books that tend to have titles like *Modern Creationism, Biblical Creationism, Science and the Bible, Evolution: The Fossils Say No,* and *Scientific Creationism.* Such assumptions are certainly understandable because many creationist books contain sections that discuss the biblical justification for creationist beliefs. When they learned that such writings were not taken from publications readily recognized as overtly creationist but rather were excerpted from popular study Bibles, they were, in general, astonished. They had long thought that one good way of comforting Christian creationist students was to have them consider that the Bible can be interpreted easily to be in harmony with evolution (or at least not in conflict). Such positions are held by the majority of Christian colleges and seminaries, and by millions of lay Christians. However, when such creationist commentary appears directly in many study Bibles, particularly if it is one that the student is using, such attempts at reconciliation are quite often viewed by the student as being anti-biblical.

Of course there are study Bibles that do not contain openly anti-evolutionary commentary; the annotators of these Bibles often are not creationists. Some of these commentaries, however, make statements that many would consider to be misconceptions in the scientific realm and in most science classrooms. For example, the popular *Quest Study Bible,* published jointly by the large companies Zondervan Publishing House and Christianity Today, Inc., asserts that the book of Genesis "tells about many beginnings: the first plants and animals; the first man and woman."[11] When the annotators discuss Noah's flood, they state "the case is strong that this flood covered the whole world."[12] The annotators also discuss the possibility that it may have been a local flood, but using the phrase "the case is strong" favors the view that the flood covered the whole world. Nevertheless, the annotators do not discuss here the possibility that Noah's flood may have been a metaphor. Furthermore, all their points for the competing views are biblical justifications—not geological. Whether intentional or not, the implicit message is that the primary or, indeed, the only evidence to contemplate is the biblical record regarding whether this physical event was worldwide.

One might also object to this point because of this study Bible's approach. Noah's flood is a biblical account; it may seem reasonable therefore, to limit the discussion of this event only to biblically derived information. However, when sea cows are mentioned in the book of Exodus, for example, the information given in the commentary is similar to that found in science books: "This large, aquatic mammal (8 to 15 feet long and up to 1,500 pounds) was related to the manatee and could be found in the Red Sea."[13] This information is not typical fare for most theological discussions. The inclusion of such references to scientific information creates a problem: The reader may not recognize a demarcation between biblical commentary and scientific evidence on the purported physical events. Imagine the potential difficulty when a student reading this study Bible comes to believe that the biblical evidence is, in fact, strong that the flood covered the entire world and then winds up in a science class encountering strong physical evidence to the contrary.

It is not surprising that a study Bible is very appealing to many who purchase a Bible, given the convenience of having both biblical text and supposedly helpful explanatory comments between the same two covers. So for some young people who reject evolution for religious reasons, their rejection may not be engendered necessarily by an upbringing in a creationist home or in a creationist church, or as a result of reading overtly creationist publications. Their rejection of evolution could very possibly result from occasional readings of the only religious material they may own, a study Bible with a stealth anti-evolutionary message.

When Science and Scripture Disagree

When it comes to literalist beliefs concerning science and the Bible, literalists hold the belief that the Bible is inerrant. Henry Morris writes that

> We can be confident that the scientific data will correlate with Scripture all right, because the same God who wrote the Word made the world![14]

Then how do literalists reconcile the overwhelming amount of scientific data that does not correlate with literalist readings of the Bible that all living things on earth appeared in four 24-hour days less than 10,000 years ago? The answer to this question is the subject of most creationist publications. Literalists believe that scientists' conclusions must be inaccurate, a priori, whenever they conclude (after rigorous and systematic examination of data) that the cause of some phenomenon is other than what the literalists' reading of the Bible would have them conclude.

This belief may sound inconsistent with the fact that the vast majority of leading literalist creationists active in fighting against the teaching of evolution usually have science degrees from secular institutions and are typically attempting to promote a debate based on science. These individuals generally state that they have a very high regard for most aspects of science. However, they have a particular name for their kind of science (a creationist term of art, if you will). They call their science "good science" or "true science," names that certainly are not recognized by the scientific community.

> The data of true science . . . must agree with the testimony of Scripture.[15]

Good science is not defined by its methods but by whether it is in agreement with literalist readings of the Bible. So for science instructors to determine whether they are teaching good science, they must be knowledgeable in what the Bible has to say about the subject in question. For example, if after years of studying the physical evidence, a scientist or the scientific community concludes that there never has been a worldwide flood or even a flood covering a hemisphere, then there must be something wrong with their science, because the Bible clearly indicates to literalists that the earth was once covered by a flood—the one for which Noah built his ark.

This literalist belief in good science is one reason why attempts of well-intentioned science instructors sometimes fail to lessen students' resistance to evolution by explaining that science is just one way of knowing. Their explanation may include the idea that science is only a procedure or structure that helps simplify information, or the idea that students do not have to believe the science but just should understand how it works. However, these explanations often do not succeed, because many creationist students believe that what their instructors are teaching is bad science or false science. Professional creationists contend that it is the science instructors who teach evolution who are the ones who do not understand the true methods and facts of science and, therefore, must be further educated. Further educated by whom? By the creationists of course.

But which type of creationist could "properly educate" scientists—literalist, progressive, or theist? Literalists would complain that the progressive and theistic creationists misinterpret what the Bible has to say about science, and, of course, the progressives and theists would have the same argument toward each other and against literalists. Because most classroom confrontations originate from literalists, we will again focus here on the education argument from a literalist viewpoint, while providing a counterargument.

Argument and Counterargument

Literalists argue that creationists must educate science instructors (and scientists) in good (literalist) science. Institute for Creation Research (ICR) president John Morris has written that

> In the scientific realm, one must interpret all scientific observations in light of Scripture.[16]

Of course, there are problems inherent with such "education," and creationist students can understand these problems if they are explained. One of the primary problems of educating science instructors in "good science" is that the scientific community would have to recognize the Bible (as opposed to other religious writings) as the utmost authority when considering which conclusions to publish in scientific journals. Additionally, those who review and edit scientific journals would have to know the proper interpretation of the Scriptures. Since there are myriad official interpretations of the Bible even within Protestant Christianity, the literalists would necessarily have to educate science instructors because they know the correct biblical exegesis. Future science instructors would have to be trained extensively in biblical theology. After all, literalist creationists think that such training is far more important than any science or math course because the ultimate arbitrator of scientific accuracy is a correct theology (provided the matter being considered has scriptural relevance as perceived by the literalists).

Often, we have found that if we lead creationists through such an argument to its logical end, they are uncomfortable with their argument. Lay literalist creationists who attempt to "enlighten" instructors during or after class, generally are reasonable and do not want public higher educational institutions to require literalist Christian theology as part of science course requirements. In addition to acknowledging the inappropriateness of such requirements, literalist creationists would not want courses in Christian theology to be taught by instructors who do not share their particular interpretation of Scripture, or far worse, by instructors who are not Christian. Literalist Christians have another problem with requiring undergraduates to take Christian theology courses as part of their science requirements: what to do about the students who performed well in these courses but who would not accept the Bible as an authority in science. These students may even be Christians who performed in the top of their classes in both the science and theology courses, yet concluded that it would be unscientific to interpret their scientific observations in light of Scripture.

At least one creationist organization has apparently solved such a problem with regard to noncreationist institutions and instructors teaching theology. The ICR has an accredited graduate school, which offers M.S. degree programs in the fields of biology, geology, physics, and science education. All these programs have been approved by the State of California and are accredited by the Transnational Association of Christian Colleges and Schools.

The ICR graduate school states that it admits "students of any gender, race, color, national or ethnic origin to all the rights, privileges, programs, and activities available to students." However, more fully it states that it admits "*Christian* students of any gender, race, color, national or ethnic origin to all the rights, privileges, programs, and activities available to students" [italics added].[17] But simply requiring that their students be Christian is far from sufficient; they also want students "who are in full agreement with the ICR tenets."[18] The tenets are divided under the headings of biblical creationism and scientific creationism (the latter will be discussed in the following section). Excerpts from the seven biblical creationism tenets include:

> These writings [the Bible] . . . are infallible and completely authoritative on all matters with which they deal, free from error of any sort, scientific and historical as well as moral and theological.

> All things in the universe were created and made by God in six literal days of the creation week described in Genesis. . . . All theories of origins or development which involve evolution in any form are false.

The list of tenets, also called a *statement of faith,* is to be adhered to by both the institution's faculty and its students. So one way that an educational institution can insure that their science graduates will practice and teach good or true science is to make sure that they are in agreement with their Christian literalist creationist theology before admission. However, it appears that the creationists have a significant science educational dilemma:

1. Creationists contend that undergraduate science instructors who teach students that evolution is accurate are incorrect.

2. Creationists contend these science instructors are incorrect because the science they are teaching is not true science.

3. Creationists contend that true science is not taught at most secular undergraduate colleges.

4. Creationists contend that good and true science is taught in ICR's graduate school.

5. To be admitted into the ICR graduate school one must believe that evolution is inaccurate.

6. See No. 1 above.

So how do students learn good and true science? It is analogous to the complaint, often echoed by students, that to get the job they want they need relevant experience, but to get that relevant experience they need the job. The leading professional creationists complain regularly that the nation's noncreationist science instructors do not understand, but these same leading creationists apparently will not admit such instructors into their graduate programs precisely because they do not understand.

This is not an isolated ICR policy; conservative Christian graduate schools and seminaries throughout the United States have statements of faith that students must attest to before being admitted. The rationale for such exclusive enrollment is often that these institutions' missions are to further the education of those students who already share their basic doctrine, not to proselytize students to that doctrine. They see it as a characteristic of higher education: Before a student can enter into higher learning at the graduate school, he or she must first understand the basics, and the basics are not taught at the graduate school.

A solution to this situation seems obvious to those who have been through higher education. Simply have Christian noncreationist science instructors get their basics from conservative Christian undergraduate institutions. Presumably, upon graduation, they would have the creationist understanding of true science and could enroll in creationist graduate schools such as ICR. However, the same problem exists at this level as it does at the graduate level: Conservative Christian undergraduate institutions typically have statements of faith to which incoming students must attest or incoming students must be Christians.

The statements of faith of conservative Christian undergraduate institutions typically do not mention creation/evolution or science. They remain rather neutral on this subject not only in their statements of faith but also in their teaching of the subject.[19] Many of these conservative Christian undergraduate schools present multiple positions, such as the literalist, progressive, and theistic creationisms, without favoring one over the other. Their students are encouraged to choose whichever position best fits with their beliefs. Even many of their faculty members differ among their beliefs on the subject within the same school. This presents a problem even for those who choose to believe a progressive or theistic creationism (or for Christian noncreationists), because it puts them all in disagreement with the ICR enrollment tenets, thus derailing any plans to learn true science. So by literalist creationist standards, these students will be teaching and practicing something other than true science, despite graduating from a conservative Christian undergraduate school.

That being the scenario for many Christian students, the "plight" is far worse for the non-Christian applicant. The non-Christian who wishes to be a future science instructor would not even meet the eligibility requirements at many conservative undergraduate Christian schools because they are not Christians. Applicants to such schools can read in the school's application information or on their Web sites that they must be willing to claim they are Christians, and, at some large schools, applicants must have been Christian for at least one year and submit references from outside their own family that confirm their Christian testimony.[20] So the non-Christians may go to a secular college and be forever "doomed" to teach something other than good and true science.

Due to these situations, future science instructors cannot be admitted to the "proper" institutions to learn about true science unless they already have some understanding about this subject. Now consider the problems encountered when one tries to solve the dilemma. In short:

1. Creationists contend that what separates true science from false science is its agreement with an accurate reading/interpretation of the Bible.

2. Creationists contend that an accurate reading/interpretation of the Bible is a Christian literalist creationist reading/interpretation and no other.

3. The place to get a college education in Christian literalist theology is a conservative Christian college.

4. Conservative Christian undergraduate schools commonly do not allow non-Christians to be admitted.

There appear to be multiple blockades erected to keep science instructors who teach "false" science from getting the education the creationists contend they so desperately need. However, there is a way they can obtain this education—a way that seems blatantly obvious to many

Christian creationists. The future science instructor should first become a born-again Christian. However, desired admittance to a conservative Christian college for training as a science instructor is certainly not a reason, in Christian creationists' minds, for a person to become a Christian. They would contend that becoming a Christian is the most important thing in any human's existence and is not undertaken merely to acquire a job. Moreover, many conservative Christians will state that a Christian college is a place for Christians to grow in their faith and knowledge about the Bible, and is not really an educational setting for non-Christians. As shocking as it may sound, the logical end to the creationist argument is that for science instructors to be properly trained to teach good or true science, they have few options other than to become a born-again Christian prior to college, attend a conservative Christian college that at least presents literalist creationism as one option, and accept that literalist creationist position as being accurate so they can distinguish between good science and bad science. While it is probably safe to say that lay Christian creationists would want all science instructors to be Christians, most would not hold to the position that science instructors be *required* to go through such a theologically narrow educational system.

This counterargument is a useful one to illustrate to lay creationists the logical result of their argument. After being presented with this counterargument, most seem to realize that giving science instructors such litmus tests is unreasonable. Most of us recognize that if students understood philosophy of science better than they do, then there would be fewer clashes between religious ideologies and science.

Creationist Philosophy

The only true philosophy . . . must be Christ-centered—which means centered in the true God and Creator of all things. . . . This would absolutely preclude any form of evolutionism or pantheism or humanism.

—Henry Morris[21]

Morris is not just opposing a philosophical position based on evolution here; philosophy in general is not considered a good thing in the Bible by some believers. The terms *philosophy* and *philosophers* are used only once each in the Bible and both times in disapproving manners. "Beware lest any man spoil you through philosophy and vain deceit, after the tradition of men, after the rudiments of the world, and not after Christ" (Colossians 2:8). The typical Christian commentary on this verse is simply that any philosophy not based on Jesus is essentially evil and potentially bad for Christians.

The only other verse that mentions philosophy concerns one of Jesus' disciples attempting to preach in Athens: "Then certain philosophers of the Epicureans, and of the Stoicks, encountered him [Paul]. And some said, What will this babbler say? other some, He seemeth to be a setter forth of strange gods: because he preached unto them Jesus, and the resurrection" (Acts 17:18). The philosophers here are reported to have called God's apostle Paul—considered a hero in Christendom—an unfavorable name apparently because he was preaching ideas that were about Jesus and His resurrection from the dead.

Whether as a message or messenger, philosophy is seen by some as being characterized as bad in the Bible. Many non-Christian science instructors we have talked with over the years have told us of their attempts at explaining to their creationist students and other related parties their philosophy of how one can accommodate evolution, science in general, and theology. They usually report less than satisfactory results. Unfortunately, many creationists just dismiss these potential solutions out-of-hand because of the association with what they perceive as non-biblically based philosophies.

4

Why Students Reject Evolution: Nonreligious Reasons

"The laws of science demonstrate that evolution is impossible, and the fossil record demonstrates that it never occurred even if it *were* possible. Evolution is nothing but a naive and credulous religious faith in the omnipotence of matter—a faith exercised blindly, in spite of the universal evidence against it."

—Henry & John Morris, ICR[1]

"I am convinced that the integrity of science education in the United States and abroad is directly threatened by such [creationist] nonsense."

—Niles Eldredge, Curator, Invertebrate Paleontology, American Museum of Natural History[2]

A great number of students think evolution is inaccurate not solely for religious reasons but for a combination of religious and nonreligious reasons. Quite often their nonreligious reasons for rejecting evolution are related to their religious beliefs. The professional literalist organizations certainly understand this connection and use many related theological and nontheological approaches to convert progressives and theists to a literalist position. Likewise, progressives use similar tactics in an attempt to convert literalists and theists.

Yet many science instructors are under the impression that the entire phenomenon of rejecting evolution is solely a religious issue, and they are quite surprised when confronted with what often seem to be nonreligious challenges to what they are teaching about evolution. These nonreligious rationales are primarily misunderstandings concerning science content or process, and are usually some of the issues discussed in creationist publications, on creationist speaking tours, and during publicly held evolution/creation debates. Many of these misconceptions (not considered misconceptions by the professional creationists) are also propagated as "good" or "true" science by literalist organizations. Such conceptions held by students are important for instructors to understand. A review of the research literature on alternative conceptions in science concludes that

> Learners bring a diverse array of ideas about natural objects and events to their science classes, and these ideas are often at variance with the scientifically accepted views in physics, biology, and chemistry. Without knowledge of these student ideas, the science teacher is at a great disadvantage.[3]

It is strongly recommended that science instructors access their students' prior knowledge concerning these nonreligious misconceptions to better address them pedagogically in the classroom. The following section presents a short discussion of some of the professional creationist nonreligious arguments that are often likely candidates for students to bring to their science courses. However, the discussion given here is only a brief overview of the typical misconceptions. If students or others bring to the science classroom more in-depth challenges that are unfamiliar to instructors (often courtesy of professional creationists), then consulting some of the resources in the Appendices may be helpful.

Characterization of Science

Creationists typically have their own characterizations of science that are not shared by the vast majority of scientists or science instructors. Almost all types of creationists want the scientific community to recognize, *as part of science*, some form of supernatural or intelligent extraterrestrial causation. Whether literalists, progressives, theists, or intelligent design advocates, creationists' arguments are quite similar on this topic: Science should not seek only naturalistic explanations but, when "necessary," should permit nonnaturalistic explanations as legitimate scientific theory. Creationists contend that methodological naturalism puts blinders on scientists who would otherwise conclude that empirical evidence points toward intelligent design. For example, some concentrate on arguing that molecular biology contains many biochemical machines that are irreducibly complex; therefore, science should conclude that they have been designed.

Modern science, being limited to investigating natural phenomena and searching for natural causes, cannot speak to whether supernatural beings exist. It seems creationists have found an insurmountable task in attempting to convince the scientific community of the so-called efficacy of using intelligent design explanations in science. Therefore, it appears that much emphasis on spreading this word is being placed on science education.

Another argument that many prominent creationists put forth is that evolution should not be considered a science because we cannot "see" it (e.g., marine to terrestrial evolution) currently occurring in a short period of time in front of our eyes. Creationists believe that if something cannot be observed occurring then it is not, as they often put it, real science. Of course this extremely restrictive definition of science would eliminate great portions of science as we know it, but, in many cases, creationists are more than ready to eliminate many of those portions. (After all, numerous portions support evolution in one way or another.)

Many creationists also believe that for a theory to be scientific it must be capable of a test of falsification as embraced by the works of philosopher Karl Popper. In other words, they attempt to apply, by name, Popper's falsification criterion whereby there must be a test that can be conducted to show a theory false, if it is false. They use this approach in attempting to show that the theory of evolution is not capable of a test and therefore should not be considered a scientific theory. (They choose to refer to evolution as a scientific model or religious theory.) What creationists fail to accept is that philosophers and historians of science have attempted for decades to illustrate to all who would listen that such a requirement would either discount most of what we currently call science (evolutionarily based or not) or would allow virtually any theory to count as science.[4]

For example, prominent philosopher of science Larry Laudan writes that there is "a host" of "reasons familiar in the philosophical literature [why] neither verificationism nor falsificationism offers much promise of drawing a useful distinction between the scientific and the non-scientific."[5] Philip Kitcher, Editor-in-Chief of *Philosophy of Science* (a leading journal in its field), holds that "The falsifiability criterion adopted from Popper . . . is hopelessly flawed."[6] Ernst Mayr, considered by many to be the world's greatest living evolutionary biologist, contends that "It [falsification] is particularly ill-suited for the testing of probabilistic theories, which include most theories in biology. . . . And in fields such as evolutionary biology . . . it is often very difficult, if not impossible, to decisively falsify an invalid theory."[7] Even introductory philosophy of science textbooks discuss the demise of falsification.[8] Therefore, by using this demarcation criterion advocated by creationists, many or most biological theories would not be considered science!

Unfortunately, much of this discussion of the demise Popper's falsification test has not reached (for various reasons) science and science education books and articles, and has not achieved widespread understanding in the science and science education communities. So as the creationists contend that Popper's falsification is a criterion of science theory, it may be possible that many science instructors agree with those who question evolution on that point. That being the case, we should not be surprised when some creationist students and the lay public ask questions concerning such falsifiability of evolutionary theory. Rather, we should be prepared to help them learn why philosophers and scientists consider it not applicable.[9]

However, despite the more philosophically inclined problems concerning Popper's falsification, many scientists freely contend that evolution could be falsified in many ways for them personally. For example, the existence of a genuine Cambrian human would do it for many. In this regard, there are potential discoveries that would cause scientists to reject evolution. However, if an equally devastating find is made against creationism (which has occurred many times), it is merely attributed to an act of God's indiscernible will.

Theory and Law

Creationists mislead students as to the meaning of the words *scientific theory*, and, in so doing, attempt to demean the status of evolution as taught in the science classroom. Many creationists tell students that the difference between a theory and a law is that a theory is merely a hypothesis that has been tested successfully numerous times. A law is nearly (or is) a scientific fact because it has been successfully tested many times.[10] After establishing these definitions the students are encouraged to consider that evolution is not a law but *only* a theory, therefore setting evolution into a supposedly inferior category than if it were a law. Because evolution is only a theory—not something that is considered factual in the scientific community—opposing theories should also be presented, such as scientific creationism. Understandably, to many students (and others), the argument sounds quite compelling.

This unfortunate characterization of the scientific use of the words *theory* and *law* has led students to believe that if scientific theories had enough evidence and were tested sufficiently with resulting supporting outcomes, then scientific theories would become scientific laws. Even many of us in science were taught these definitions in our public school science classes. As such, it is certainly understandable why these misconceptions are common among students.

These misconceptions of the meanings of the words *theory* and *law* are reinforced by the meaning of the word *theory* as something that is not factual—as merely a guess without any or with very little supporting evidence. This use of the word is so ingrained in our culture that even many scientists, when casually speaking, will use the term *theory* in the same way that the general public uses the word. However, when scientists communicate in a more scientifically exacting manner in research journals, they normally use it to mean an explanation of phenomena that have been rigorously tested, and they use the word *law* to mean a descriptive generalization of phenomena.[11] No matter how extensively a theory is tested and despite incontrovertible data supporting its accuracy, a theory does not become a law (i.e., explanations do not become descriptions).

These conceptions of theories and laws are very difficult for most students to learn because their prior conception of the word *theory* is so entrenched. Even short-term classroom activities designed to confront students' misconceptions will probably not facilitate change. However, some success may occur if all science instructors in all science courses consistently used the terminology properly and constructed activities that challenged students to wrestle with the scientific versus the popular use of the word *theory*.

"Horizontal" *vs.* "Vertical" Evolution

An understanding of evolution is important when trying to control mutating pathogens, pests, and infectious diseases. Certainly, most science instructors would agree with such a statement. Surprisingly to many who teach evolution, so would many creationists—with the addition of just one word. However, it is the meaning of that one word (the word *horizontal*) that helps engender evolutionary misconceptions.

Although creationists don't like the word *evolution,* they are willing to use it when referring to changes among organisms that, in their view, are merely changes within the same "kinds" of organisms (that is, cumulative changes resulting in descendants that are roughly the same; e.g., dogs into other types of dogs—their example). Creationists call such change *horizontal evolution.* They say that horizontal evolution does not constitute an increase in complexity and therefore should not be confused with the more global meaning of biological evolution—*vertical evolution*.

Creationists begrudgingly would like to see the label *horizontal evolution* used instead of the term *microevolution* (not that the two are synonymous). Their greater preference, however, is use of terms such as *horizontal changes, horizontal variation,* or *variation within a kind.* Creationists take the word *kind* from Genesis ("each unto his own kind"). Even literalists contend that organisms *do* have the ability to change over time within certain limits (within kind) because of environmental pressures, although the exact limits usually are not specified in creationist literature. A few leading creationists believe these limits probably to be approximately at the phylogenic level of family.[12]

Therefore, many creationists are not concerned that science instructors discuss changes over time that result in differing species, genera, and sometimes families because these changes are merely changes within a kind. They do not view such change as a type of evolution that increases in complexity and, therefore, typically have no problem with statements such as "an understanding of evolution is important when trying to control mutating pathogens, pests, and infectious diseases." Creationists simply see the term *evolution* used in this context as meaning change over time within a kind. Consequently, many creationist students who hold this interpretation of evolution find instruction about organisms evolving into different species or genera totally acceptable.

Nevertheless, when evolution instruction reaches beyond the level of family, creationists quickly find such teaching to be problematic. Therefore, many of the examples of speciation that science instructors use in class, although scientifically illustrative, may be irrelevant to creationist students. Likewise the term *natural selection* may be unacceptable to them when used in contexts resulting in cumulative change beyond a kind. In this context, then, it is important to teach students that even if natural selection were disconfirmed, the theory of evolution would not therefore be disconfirmed. Students need to learn the difference between evidence for the occurrence of evolution and proposed mechanisms for the occurrence (e.g., anatomical homologies, biochemical homologies, the fossil record, etc.).

We end this section with a warning: Creationists usually have a different meaning for the word *macroevolution* than its usual scientific meaning, evolution of taxa higher than the level of species. Creationists typically use the word *macroevolution* to mean anything above the created "kinds"—large-scale evolutionary change. So microevolution is most often acceptable to creationists while macroevolution (by their definition) is not acceptable. Be aware that creationist students do not automatically share the same understanding of a word as accepted in the scientific community just because they are using the word in their arguments.

"Missing Links"

Many creationist students are convinced that no fossil evidence shows any cumulative effect of evolution beyond "kinds." Although instruction in the fossil record is important in evolution education, many biology instructors say they spend no time on paleontology despite spending significant amounts of time on other evidence for evolution. Many instructors think that paleontology is more appropriately taught in geology courses than in biology courses. Yet most students never take geology, although many take biology. That biology course may be their only opportunity in their formal education to learn about the fossil evidence for evolution.

Despite the fact that many creationist students are educated in public institutions, much (if not all) of their ideas regarding paleontology come from creationist publications and presentations. Creationist leaders profess that no transitional sequences exist to which science instructors can point, and they believe that this matter is crucial in attacking evolution.

> The data from the fossil record do not agree with the predictions of the evolution model. Therefore, whether or not the earth is ten thousand, ten million, or ten billion years old, the fossil record does not support the general theory of evolution.[13]

This statement was written by Duane Gish, Senior Vice President, ICR, in his very popular book among creationists titled *Evolution: The Fossils Still Say No!* His previous book by the same title (*sans* still) was so popular that it has gone through numerous editions and printings over the last few decades. He wrote this book in the hopes of convincing others through the use of his interpretation of the fossil record (he is trained as a biochemist), that there are no transitional forms and therefore evolution did not occur. (Contrarily, as a general principle, students should learn that absence of evidence is not necessarily evidence of absence, and should be taught the nature of the fossil record that leads paleontologists to conclude that evolution did occur.)

Beyond using his own fossil record interpretations as evidence that evolution did not occur, Gish employs another tactic of using others' statements. He selects quotes from a variety of distinguished paleontologists and strings the quotes together with his commentary in between. His goal in doing so is to build an argument that would compel the reader to conclude that the paleontological evidence for evolution is nonexistent. His method is somewhat analogous to collecting selected quotes of criticisms of the university system from members of university faculties and making them collectively sound like the university system has failed. Separately, such criticisms are generally meant to be constructive and to improve the operation of universities. However, taken out of context and strung together in a particular order with anti-university commentary interjected in between, these criticisms could be made to appear that faculty members think all university systems are worthless and should be shut down.

Of all the living authors Gish cites in his nearly 400-page book—whether scientists, creationists, theologians, or others—the person he cites most (on 17 different pages) is Robert Carroll of McGill University in Montreal. Carroll is an eminent paleontologist with a highly distinguished career. He is also author of numerous books on evolution including the most widely adopted vertebrate paleontology textbook in the world.[14] Carroll and other renowned paleontologists who have studied the fossil record for decades are compelled by the overwhelming evidence to conclude that evolution did occur. However, by Gish's tactics noted above, students often believe otherwise.

Gish is not alone in using these methods. This tactic has been so successful with the general public and to some extent with students, that the president of ICR has published a nearly 500-page book containing essentially nothing but quotes from evolutionists. *That Their Words May be Used Against Them* has the following printed on the cover: "Quotes from Evolutionists Useful for Creationists."[15] The title of the book communicates its intended use: as ammunition against those who claim that the scientific evidence points toward evolution.

When students read or hear the out-of-context quotes from such a book, they quite often think that evolution must be a theory in crisis within the scientific community. If creationist students suddenly quote respected evolutionary scientists and expect science instructors to respond instantly (without even knowing the context of the quotes) we recommend the following two actions: (1) instructors should explain to the students that they are cognizant of the scientists being quoted (if they are) and that the evidence compels these scientists to conclude that evolution is an accurate scientific theory, and (2) instructors should request that the students bring in the *original sources* of the quotes so that they can be read in context. Having the original source (not the quote book) will allow instructors to illustrate to students that there is an entire book or article surrounding the one quote and that the publication is not challenging the occurrence of evolution. The goal of this exercise is not for students to find evolution compelling simply because experts find the data compelling; they should examine the evidence and come to their own conclusions. However, when students quote these experts, they need to understand clearly the positions of the people they are quoting and the contextual meaning of the quotes.[16]

Classroom activities on fossil lineages can be useful to help students examine the evidence for evolution so that they can come to their own conclusions. For example, provide students with some well-documented fossil records (e.g., a lineage with representative members of 55 million years of the elephant lineage[17]) and challenge them to provide a scientific explanation, other than descent with modification, for the temporal (sequential) appearance of these morphologically similar organisms. Why do they change over time?[18] After presenting nonhuman organisms, many instructors find it useful to then introduce some hominid transitions (e.g., transitions from something rather apelike to modern human beings over the last 3 million years: *Australopithecus africanus,* brain size ~450 ml; *Homo habilis,* ~750 ml; *Homo ergaster,* ~1,000 ml; *Homo sapiens,* ~1,350 ml.[19] The goal of such an exercise is for students to confront their misconceptions concerning paleontological evidence for evolution.

Regardless of whether students overtly challenge the presence of fossil transitions, the fossil record should be addressed in introductory lessons about evolution to confront students' common paleontological misconceptions. Students should learn about the abundance of many intermediary fossils, including land mammals to cetaceans, reptiles to mammals, dinosaurs to birds, and primitive hominids to *Homo sapiens.* This will help prepare students to answer questions outside academia concerning such topics. In this way, when students hear creationist-type claims of a lack of paleontological evidence for evolution, they will be better prepared with scientific data and rationale to consider such unscientific claims. Understanding the nature of the fossil record is an essential part of basic scientific literacy.

Punctuated Equilibrium

Since the 1970s, one idea stands out in evolutionary science as having been the most contorted by leading creationists in an attempt to further their views. This idea, the theory of punctuated equilibrium, is therefore widely misunderstood by creationist students. In 1972, paleontologists Niles Eldredge and Stephen Jay Gould shook up the scientific world with the publication *Punctuated Equilibrium: An Alternative to Phyletic Gradualism*[20] and the theory's presentation at a conference on microevolution at Chicago's Field Museum of Natural History in 1980. It continues to be a significant topic of discussion in many scientific disciplines.

In essence, phyletic gradualism involves species developing gradually, leaving numerous transitional forms, and the entire geographic population generally speciates. In contrast, punctuated equilibrium involves geologically rapid speciation with long periods of stasis, therefore leaving relatively few transitional fossil forms; speciation occurs in only a small subpopulation of the entire geographic population. In part, Eldredge and Gould developed punctuated equilibrium to help explain the appearance of portions of the fossil record that included long periods of unchanged fossils followed rapidly by new types of fossils. It is with this point of rapidity that creationists have distorted the science.

Professional creationists believe (and many students hold the misconception) that punctuated equilibrium explains the appearance of distinct kinds of organisms by a rate so fast that a single generation would be sufficient for the punctuation to occur. The creationists like to represent the process of punctuated equilibrium as a bird hatching out of a reptile's egg, and, unfortunately, such images often remain in the minds of many. To understand punctuated equilibrium accurately, students must learn that the "punctuation" in punctuated equilibrium may constitute hundreds or thousands of generations, yet such stretches of time represent the blink of an eye in a geological sense. A "geological blink" does *not* mean a time frame in which a bird hatches from a reptile egg, as in the creationist characterization.

Creationists also try to paint the picture that punctuated equilibrium means that there are no transitional forms to be discovered in the fossil record. Obviously, this is not what punctuated equilibrium predicts. A punctuational event, which occurs in a small isolated subpopulation, can take 50,000 years to occur and can leave hundreds of thousands or even millions of fossils. In comparison to the long periods of stasis of the entire ancestral form, the relatively short period of punctuation yields a relatively low percentage of intermediate fossils, which are therefore more difficult for paleontologists to discover than fossils that are more bountiful.

Contrasting the tempo and mode of phyletic gradualism with punctuated equilibrium is a vital part of learning evolution, and students should understand reasonably well what punctuated equilibrium explains and predicts.

Dinosaurs and Human Tracks

Early in the 1900s, some dinosaur and supposed fossilized human tracks were discovered coexisting in very close proximity (in some cases overlapping in the same sediment layer) in the Paluxy River bed, near Glen Rose, Texas. Many decades went by without much publicity until the late 1960s, when significant attention by the creationists was focused on the tracks. Creationists occasionally have put forth other human footprints (often purported as giant human footprints) as existing with dinosaur tracks in Cretaceous strata, but the Paluxy River tracks became a focal point for numerous creationists.

Through most of the 1970s and 1980s, many creationists concluded that the Paluxy tracks were a critical if not fatal blow to evolutionary theory. They contended that the tracks illustrated clearly that humans and dinosaurs coexisted despite paleontologists' assertions that dinosaurs became extinct long before humans evolved. Creationists used these "data" in publications, lectures, debates, radio programs, and films to sway all who would listen into believing that evolutionary scientists were wrong about evolutionary theory.

One of the most popular and seemingly influential publications in creationist history is a very small booklet (also called a tract) entitled *Have You Been Brainwashed?* that has been distributed to thousands of students at educational institutions across North America. The booklet, authored by Duane Gish, states that

> Near Glen Rose, Texas, fine clear tracks of dinosaurs and man have been found in the same rock formation. These prints are within a few yards of each other, and sometimes even cross each other.[21]

This "find" was even used in one of the most influential creationist publications of all time, the early 1960s book *The Genesis Flood*. In the mid-1970s, the Paluxy River tracks appeared in ICR's widely circulated periodical *Acts and Facts.* The tracks were given eight small-print pages in the insert "Impact Series: Vital Articles on Science/Creation."

> Thus, evolutionists face a dilemma. There is, at the very least, a 60-million-year gap between the age of the dinosaur and the advent of man, a gap during which there should be neither man nor dinosaur. But in Glen Rose, the fossils indicate that man and dinosaur lived not only at the same time but even at the same place![22]

Then, in the early 1980s, ICR gave full front page coverage in *Acts and Facts* to an entire book on the on the Paluxy footprints authored by John Morris: *Tracking Those Incredible Dinosaurs: And the People Who Knew Them. Acts and Facts* stated:

> The famous overlapping dinosaur and human footprints in the Cretaceous Glen Rose formation of Texas are documented and discussed in detail. . . . Dr. Morris has made numerous field studies at the site of these remarkable footprints, extending over a period of eight years and has also assimilated and analyzed all studies of these tracks by other investigators. The book is documented with over 200 photographs, providing key evidence for the contemporaneity of man and dinosaur.[23]

The book was an attempt to document the evidence of the tracks and those involved in seeing the tracks firsthand in the hopes that people reading it would conclude, as the author Morris did:

> The evidence is extensive and that it demands the conclusion that man and dinosaur walked at the **same time** and in the **same place**.[24]

So for decades, large numbers of students and others were led to believe that, contrary to evolutionary theory, humans coexisted with dinosaurs. Of course, numerous paleontologists have examined the exhibits and visited the site and have concluded that no evidence exists to support the idea that humans and dinosaurs lived at the same time.[25] (The supposed human tracks are false while the dinosaur tracks are authentic.) In other words, the Texas "footprints" definitively have been debunked. Subsequently in 1985, the creationists withdrew their book from the market and in 1996 John Morris wrote with regard to the tracks:

> I am of the opinion that the evidence is, at best, ambiguous and unusable as an anti-evolutionary argument at the present time.[26]

Even though many creationist organizations no longer use these tracks as evidence for the coexistence of humans and dinosaurs, the myth is difficult to extinguish. It continues to linger among many lay-creationists who are still unaware that their own leaders have discarded the tracks as useful. Shockingly, an occasional television or radio program is aired discussing how scientists are puzzled over the tracks. In 1995, for example, NBC broadcast a television program titled *Mysterious Origins of Man* in which the Paluxy footprints were discussed as if they had not been debunked, contributing once again to the belief that physical evidence exists for human/dinosaur coexistence. This particular media episode inspired anthropologist John Cole to question:"Does NBC have NO fact-checkers??!!"[27]

What about that popular and influential booklet *Have You Been Brainwashed?* It also has changed since the debunking of the Texas tracks, but has changed strangely. We purchased this edition of the booklet in person from ICR's museum bookstore in May, 2000. While we expected it to contain no admission of the debunking, we assumed that there would be, at the least, no mention of the tracks. We were wrong.

> Near Glen Rose, Texas, undoubted dinosaur tracks have been found. In the same strata and very close by, other tracks have been found that some claim are human, although such claims are controversial.[28]

Controversial?! These "human" tracks are only controversial within the creationist community. Within the scientific community they are a long-past debunked blip, and today are a nonissue.

Science instructors should not be surprised when students bring up the subject of these tracks, but should inform students that the tracks have been discredited. Science instructors might also inform students that today virtually no creationist organizations use the Paluxy tracks as an argument against evolution. However, it will likely be some time before the "Flintstones syndrome" disappears.

Dating Fossils and Rocks

The dating of certain fossils and rocks is not a crucial point to many creationist students. They are more concerned with the global matter of the age of the earth, which is often very important in their acceptance or rejection of evolution. No matter how well science instructors may help students learn many aspects of evolution, they will still not convince many creationist students that the earth is old enough for evolution to have taken place and this will remain the primary reason why they believe evolution is inaccurate. Over the course of their science instruction, some creationist students may come to accept that evolution sounds logical and perhaps may recognize that it could have actually occurred (in principle) . . . if only there had been enough time. Most literalist creationists believe that their reading of the Bible does not allow for the age of the earth to be greater than 10,000 years. Additionally, they feel that they have numerous nonreligious reasons for believing that the earth is very young.

Leading literalist creationists popularize the idea that scientists use only dating techniques that allow for enough time for evolution to have occurred. The creationists say that evolutionary scientists favor the use of radiometric dating techniques, which yield billions of years, while discarding other techniques that yield ages too short for evolutionary processes. Additionally, they contend that this selection of dating techniques is not based on scientific data but is arbitrary.

Although the majority of students in North America are not exposed to such misinformation, it is amazing how many noncreationist students have come to believe that dating techniques are questionable or, simply, that scientists are not sure that the earth is very old. Students tend not to recall where or why they have come to believe this, but their belief may be related to the public's general attitude that anything that happened in the past (especially the distant past) is somehow less scientifically accurate or credible than something more current. This young-earth misconception usually is not held very strongly by noncreationist students. The idea tends to be something that they heard somewhere but never really thought about. Therefore, instruction on dating methods is important to students' evolution literacy in general. More specifically, instruction on useful dating methods is important for creationist students who have never been taught in science courses why these techniques are sound, but who may be somewhat knowledgeable about creationist arguments against them.

Origin of Life

The great naturalist and geneticist Theodosius Dobzhansky wrote:

> Evolution is a process which has produced life from non-life, which has brought forth man from an animal, and which may conceivably continue doing remarkable things in the future.[29]

Not surprisingly, most creationists disagree with Dobzhansky on this point. Whether it be origin of life by evolution or some other natural science explanation, creationists protest. When it comes to scientific investigations into the origin of life, creationists ranging from literalists to intelligent design advocates argue that science should include supernatural (or nonsupernatural extraterrestrial) agency. Creationists also want science instructors to teach about it in public educational institutions.

Intelligent design advocate Michael Behe contends that because science currently cannot explain the origin of molecular life to his (and to others') satisfaction, the scientific community should officially recognize "that life was designed by an intelligent agent."[30] Literalists surely don't agree much with Behe, with his acceptance that the earth could be quite old, and with his belief that organisms may share a common ancestor.[31] Nevertheless, they do agree with him on molecular intelligent design. As literalists often do, Henry and John Morris write with more certainty than others about the impossibility of science explaining the origin of life by natural causes:

> Life, at the very simplest level conceivable, has absolutely no possibility of having been generated by any other means than special creation by a living Creator.[32]

Creationist nonreligious rationales for rejecting a scientific explanation for the origin of life generally come from arguments concerning:

1. *irreducible complexity*—if its complexity were reduced it wouldn't have any function,

2. *mathematical probabilities*—it's improbable,

3. *thermodynamics*—things decrease in energy/complexity, and

4. *biochemistry*—unknown pathways/mechanisms.[33]

Many students assume that if a science instructor is going to teach about biological evolution, the subject must necessarily include learning about the origin of life. (In fact, soon after Darwin disclosed his theory in 1859, many of his critics claimed its shortcoming was not explaining how life began.) There are many differing views within the educational community on where the lessons concerning evolution should begin; we will discuss two here. One view is that the lessons should include the study of prebiotic evolution. The other view is that lessons should not include prebiotic conditions but should start only after living organisms developed.

As one would expect, when topics concerning the origin of life are introduced in class, instructors typically encounter a number of students who feel that their religious beliefs are being challenged. This is the case even though most of the leaders of the major religious organizations in North America consider it fully acceptable that God may have chosen to use evolution (including prebiotic evolution) as the preferred mechanism to bring humans into existence. However, as is often the case, many of the lay-participants of such religions are unaware of the positions of their religious leaders and so believe the contrary.

This being the situation, some inventive college instructors distribute to their students copies of official statements from major religious organizations that support evolution.[34] Students, whether creationist or not, usually are surprised to learn that so many religious organizations find evolution to be compatible with their faiths. Subsequently, many students no longer think of the "controversy" as evolution versus all religions, but rather as a conflict between evolution and a small number of religious factions. This understanding usually lessens the degree of animosity students hold toward the subject and the instructor.

However, for those science instructors who teach in public high schools, using such a method could produce—and has produced—a creationist-parental backlash. Parents complain that science instructors should not be distributing religious statements but rather should stick to teaching science. They also allege that the science instructor has a hidden agenda of changing a tenet of their child's religious belief. Because of this backlash or its potential, many high school biology instructors consider it best (and appropriate) *not* to include prebiotic evolution in their teaching of biology. Aside from the backlash problem, there are many other reasons for advocating this view.

One rationale is that a discussion of prebiotic evolution might be highly offensive to some students and, therefore, might result in their wholesale rejection of biological evolution when, in fact, prebiotic evolution is probably too abstract and complex for most high school students to grasp fully anyway. Another rationale for this view is that discussions of prebiotic events are simply not appropriate for studying the *biological* theory of evolution. Advocates of this view think that prebiotic discussions are primarily the domain of chemistry and geophysics, not of biology. Therefore it is prudent for biology instructors to use their course time to concentrate exclusively on biological concepts.

Conversely, advocates of an integrated science approach often find the inclusion of prebiotic concepts useful when discussing probability, thermodynamics, and the vastness of the universe. Moreover, many adult Americans hold numerous misconceptions about these topics that contribute to their rejection of evolution. And for many secondary students in the United States, biology is the last science course that they will take in their lifetimes. Therefore, many educators advocate including such topics in high school biology courses to improve the population's scientific literacy.

Laws of Thermodynamics

The laws of thermodynamics are often an important issue to many leading creationists, with the second law being the primary focus. The typical argument attempts to show that an increase in complexity via evolution is in contradiction with the second law. However, most secondary school and college undergraduates (other than those who major in physics and chemistry) either think that the argument is not that compelling or decide not to use the argument because they do not feel confident in their understanding of it. (Of course, the earth is not a closed system but gets vast energy inputs from the sun. Even the surface of the primitive earth had many energy sources that could have been available for organic synthesis—solar radiation, electrical discharges from lightning, shock waves from meteorite impact, radioactivity, volcanoes, and cosmic rays.) Most students, whether creationist or not, profit from a short introduction to the basics of thermodynamics and how the second law does not preclude evolution from occurring. (For a review of the second law of thermodynamics in relation to evolution see Patterson 1983, and Strickberger 2000.)

Plate Tectonics

Yes, there are some prominent literalist creationists who even question that plate tectonics (earlier known as continental drift) occurs, including most, if not all, of the underlying science (including magnetic-field reversals, sea floor spreading, and the formation of mountains). Literalist creationists do not state that plate tectonics does *not* occur, but they say that they do not find the evidence compelling, so they are undecided. Naturally, any concepts inherent in the plate tectonics theory that require periods of time longer than 10,000 years would be considered unacceptable by literalists out-of-hand.

It is interesting to note the criterion used by many creation "scientists" to make their decisions, whether to be pro, con, or neutral on such a seemingly remote topic as plate tectonics. The criterion is whether it has a potential effect of supporting evolutionary theory. If it helps support evolution, then it must be rejected a priori. The decision is not based on the merits of the particular concept itself. The presidents of ICR have instructed like-minded creationists as to what position they should take on the issue and why. (The latter reason is important to science instructors who want to learn their students' rationales for rejecting seemingly evolutionary-distant scientific concepts.) Here is an example of that counsel with regard to continental drift:

In the meantime [until more evidence can be gathered that creationists find suitable in favor of continental drift], it is unnecessary for creationists to take any specific position on the subject, since it does not affect the creation/evolution issue one way or another. When it is incorporated as an integral part of the long-age evolutionary model, of course, as some have done, then it can and should be rejected as unscientific and false.[35]

It is very probable that the vast majority of creationist students in our classrooms do not make the connection between seemingly distant scientific concepts, such as plate tectonics, and their relevance (pro or con) to evolution, and therefore they do not immediately reject the evidence or logic of these concepts. However, creationist students who have learned that certain ideas support evolution are likely to reject immediately the evidence or logic of those concepts. Science instructors who have such students enrolled in their courses have an incredible teaching challenge ahead of them.

Many instructors who have de-emphasized the teaching of evolution in their courses or who have completely eliminated it (both done usually in the hope of lessening conflict in the classroom and school) are surprised when students reject other science concepts being taught (e.g., magnetic-field reversals, the spreading of the sea floor, formation of mountains, or the entire concept of plate tectonics). Some of these students hold misconceptions about a large number of related concepts in all major areas of science that support evolution. In addition, many students' rejection of evolutionary concepts is based on their misunderstanding of many methods of science in general.

5

Why Should Students Learn Evolution?

"When you combine the lack of emphasis on evolution in kindergarten through 12th grade, with the immense popularity of creationism among the public, and the industry discrediting evolution, it's easy to see why half of the population believes humans were created 10,000 years ago and lived with dinosaurs. It is by far the biggest failure of science education from top to bottom."

—Randy Moore, Editor, *The American Biology Teacher*[1]

"This is an important area of science, with particular significance for a developmental psychologist like me. Unless one has some understanding of the key notions of species, variation, natural selection, adaptation, and the like (and how these have been discovered), unless one appreciates the perennial struggle among individuals (and populations) for survival in a particular ecological niche, one cannot understand the living world of which we are a part."

—Howard Gardner, Professor,
Harvard Graduate School of Education[2]

With all of the controversy over the teaching of evolution reported in the media, with parents confronting their children's science teachers on this issue, and with students themselves confronting their instructors in high schools and colleges, would it be best—and easiest—to just delete the teaching of evolution in the classroom? Can't students attain a well-rounded background in science without learning this controversial topic? The overwhelming consensus of biologists in the scientific community is "no." Why, then, should science students learn about evolution?

A simple answer is that evolution is the basic context of all the biological sciences. Take away this context, and all that is left is disparate facts without the thread that ties them all together. Put another way, evolution is the explanatory framework, the unifying theory. It is indispensable to the study of biology, just as the atomic theory is indispensable to the study of chemistry. The characteristics and behavior of atoms and their subatomic particles form the basis of this physical science. So, too, biology can be understood fully only in an evolutionary context. In explaining how the organisms of today got to be the way they are, evolution helps make sense out of the history of life and explains relationships among species. It is a useful and often essential framework within which scientists organize and interpret observations, and make predictions about the living world.

But this simple answer is not the entire reason why students should learn evolution. There are other considerations as well. Evolutionary explanations answer key questions in the biological sciences such as why organisms across species have so many striking similarities yet are tremendously diverse. These key questions are the *why* questions of biology. Much of biology explains *how* organisms work . . . how we breathe, how fish swim, or how leopard frogs produce thousands of eggs at one time . . . but it is up to evolution to explain the why behind these mechanisms. In answering the key *why* questions of biology, evolutionary explanations become an important lens through which scientists interpret data, whether they are developmental biologists, plant physiologists, or biochemists, to mention just a few of the many foci of those who study life.

Understanding evolution also has practical considerations that affect day-to-day life. Without an understanding of natural selection, students cannot recognize and understand problems based on this process, such as insect resistance to pesticides or microbial resistance to antibiotics. In a report released in June, 2000,[3] Dr. Gro Harlem Brundtland, Director-General of the World Health Organization, stated that the world is at risk of losing drugs that control many infectious diseases because of increasing antimicrobial resistance. The report goes on to give examples, stating that 98% of strains of gonorrhea in Southeast Asia are now resistant to penicillin. Additionally, 14,000 people die each year from drug-resistant infections acquired in hospitals in the United States. And in New Delhi, India, typhoid drugs are no longer effective against this disease. Such problems face every person on our planet, and an understanding of natural selection will help students realize how important their behavior is in either contributing to or helping stem this crisis in medical progress.

Evolution not only enriches and provides a conceptual foundation for biological sciences such as ecology, genetics, developmental biology, and systematics, it provides a framework for scientific disciplines with historical aspects, such as anthropology, astronomy, geology, and paleontology. Evolution is therefore a unifying theme among many sciences, providing students with a framework by which to understand the natural world from many perspectives.

As scientists search for evolutionary explanations to the many questions of life, they develop methods and formulate concepts that are being applied in other fields, such as molecular biology, medicine, and statistics. For example, scientists studying molecular evolutionary change have developed methods to distinguish variations in gene sequences within and among species. These methods not only add to the toolbox of the molecular biologist but also will have likely applications in medicine by helping to identify variations that cause genetic diseases. In characterizing and analyzing variation, evolutionary biologists have also developed statistical methods, such as analysis of variance and path analysis, which are widely used in other fields. Thus, methods and concepts developed by evolutionary biologists have wide relevance in other fields and influence us all daily in ways we cannot realize without an understanding of this important and central idea.

Evolution is not only a powerful and wide-reaching concept among the pure and applied sciences, it also permeates other disciplines such as philosophy, psychology, literature, and the arts. Evolution by means of natural selection, articulated amidst controversy in the mid-nineteenth century, has reached the twenty-first century having had an extensive and expansive impact on human thought. An important intellectual development in the history of ideas, evolution should hold a central place in science teaching and learning.

Why is evolution the context of the biological sciences—a unifying theory?

First, how does evolution take place? A key idea is that some of the individuals within a population of organisms possess measurable changes in inheritable characteristics that favor their survival. (These characteristics can be morphological, physiological, behavioral, or biochemical.) These individuals are more likely to live to reproductive age than are individuals not possessing the favorable characteristics. These reproductively advantageous traits (called *adaptive traits* or *adaptations*) are passed on from surviving individuals to their offspring. Over time, the individuals carrying these traits will increase in numbers within the population, and the nature of the population as a whole will gradually change. This process of survival of the most reproductively fit organisms is called *natural selection*.

The process of evolutionary change explains that the organisms of today got to be the way they are, at least in part, as the result of natural selection over billions of years and even billions more generations. Organisms are related to one another, some more distantly, branching from a common ancestor long ago, and some more recently, branching from a common ancestor closer to the present day. The fact that diverse organisms have descended from common ancestors accounts for the similarities exhibited among species. Since biology is the story of life, then evolution is the story of biology and the relatedness of all life.

How do evolutionary explanations answer key questions in the biological sciences?

Evolution answers the question of the unity and similarity of life by its relatedness and shared history. But what about its diversity? And how does evolution answer other key questions in the biological sciences? What are these questions and how does evolution answer the *why* question inherent in each?

Evolution explains the diversity of life in the same way that it explains its unity. As mentioned in the preceding paragraphs, some individuals within a population of organisms possess measurable changes in inheritable characteristics that favor their survival. These adaptive traits are passed on from surviving individuals to their offspring. Over time, as populations inhabit different ecological niches, the individuals carrying adaptive traits in each population increase in numbers, and the nature of each population gradually changes. Such divergent evolution, the splitting of single species into multiple, descendant species, accounts for variation. There are different modes, or patterns, of divergence, and various reproductive isolating mechanisms that contribute to divergent evolution. However, the result is the same: Populations split from common ancestral populations and their genetic differences accumulate.

What are some other key questions in biology that are answered by evolution? One key question asks why form is adapted to function. Evolutionary theory tells us that more organisms that have parts of their anatomy (a long, slender beak, for instance) better adapted to certain functions (such as capturing food that lives deep within holes in rotting tree trunks) will live to reproductive age in greater numbers than those with less-well-adapted beaks. Therefore, the organisms with better-adapted beaks will pass on the genes for these features to greater numbers of offspring. Eventually, after numerous generations, natural selection will result in a population that has long slender beaks adapted to procuring food. Thus, anatomical, behavioral, or biochemical traits (the "forms") fit their functions because form fitting function is adaptive. But this idea leads us to yet another important question: Why do organisms have a variety of nonadaptive features that coexist amidst those that are adaptive?

During the course of evolution, traits that no longer confer a reproductive advantage do not disappear in the population unless they are reproductively disadvantageous. A population of beige beach birds that escaped predation because of protective coloration will not change coloration if this population becomes geographically isolated to a grasslands environment, unless the now useless beige coloration allows the birds to be hunted and killed more easily. In other words, if beige coloration is not a liability in the new environment, the genes that code for this trait will be passed on by all surviving birds in this grasslands niche. Even as the population of birds changes over generations, the genes for beige feathers will be retained in the population as long as this trait confers no reproductive disadvantage (and as long as mutation and genetic drift do not result in such a change).

These preceding examples do not cover all the key questions of biology (of course), but do show that such key questions are really questions about evolution and its mechanisms. Only evolutionary theory can answer the *why* questions inherent in these themes of life.

How does understanding evolution help us understand processes that affect our health and our day-to-day life? And how are evolutionary methods applied to other fields?

As mentioned earlier, without an understanding of natural selection, students cannot recognize and understand problems based on this process, such as insect resistance to pesticides or microbial resistance to antibiotics. Additionally, it is only through such understanding that scientists can hope to find solutions to these serious situations. Scientists know that the underlying cause of microbial resistance to antibiotics is improper use of these drugs. As explained in the World Health Organization report *Overcoming Antimicrobial Resistance*, in poor countries antibiotics are often used in ways that encourage the development of resistance. Unable to afford the full course of treatment, patients often take antibiotics only until their symptoms go away—killing the most susceptible microbes while allowing those more resistant to survive and reproduce. When these most resistant pathogens infect another host, antibiotics are less effective against the more resistant strains. In wealthy countries such as the United States, antibiotics are overused, being prescribed for viral diseases for which they are ineffective and being used in agriculture to treat sick animals and promote the growth of those that are well. Such misuse and overuse of antibiotics speeds the process whereby less resistant strains of bacteria are wiped out and more resistant strains flourish.[4]

In addition to developing resistance to antibiotics and other therapies, pathogens can evolve resistance to the body's natural defenses. The virulence of pathogens (the ease with which they cause disease) can also evolve rapidly. Understanding the coevolution of the human immune system and the pathogens that attack it help scientists track and predict disease outbreaks.

Understanding evolution also helps researchers understand the frequency, nature, and distribution of genetic disease. Gene frequencies in populations are affected by selection pressures, mutation, migration, and random genetic drift. Studying genetic diseases from an evolutionary standpoint helps us see that even lethal genes can remain in a population if there is a reproductive advantage in the heterozygote, as in the case of sickle-cell anemia and malaria.

Sickle-cell anemia is one of the most common genetic disorders among African Americans, having arisen in their African ancestors. It has also been observed in persons whose ancestors came from the Mediterranean basin, the Indian subcontinent, the Caribbean, and parts of Central and South America (particularly Brazil). The sickle-cell gene has persisted in these populations, even though the disease eventually kills its victims, because carriers who inherit a single defective gene are resistant to malaria. Those with the sickle-cell gene have a survival advantage in regions of the world in which malaria is prevalent, which are the regions of the ancestral populations listed previously. Although many of these peoples have since migrated from these areas, this ancestral gene still persists within their populations.

Scientists are also working to identify gene variations that cause genetic diseases. Molecular evolutionary biologists have developed methods to distinguish between variations in gene sequences that affect reproductive fitness and variations that do not. To do this, scientists analyze human DNA sequences and DNA sequences among closely related species. The Human Genome Project, a worldwide effort to map the positions of all the genes and to sequence the over 3 billion DNA base pairs of the human genome, is providing much of the data for this effort and also is allowing scientists to study the relationships between the structure of genes and the proteins they produce. (On June 26, 2000, scientists announced the completion of the "working draft" of the human genome. The working draft covers 85% of the genome's coding regions in rough form.[5])

Some diseases are caused by interaction between genes and environment (lifestyle) factors. Genetic factors may predispose a person to a disease. For example, America's number one killers, cardiovascular disease and cancer, have both genetic and environmental causes. However, the complex interplay between genes and environmental factors in the development of these diseases makes it difficult for scientists to study the genetics of these diseases. Nevertheless, using evolutionary principles and approaches, scientists have developed a technique called *gene tree analysis* to discover genetic markers that are predictive of certain diseases. (Genetic markers are pairs of alleles whose inheritance can be traced through a pedigree [family tree].) Analyses of gene trees can help medical researchers identify the mutations in genes that cause certain diseases. This knowledge helps medical researchers understand the cause of the diseases to which these genes are linked and can help them develop treatments for such illnesses.

How is evolution indispensable to the subdisciplines of biology and how does it enrich them?

Organizing life, for example, a process on which Linnaeus worked as he grouped organisms by morphological characteristics, continues today with processes that reflect evolutionary relationships. Systematics, the branch of biology that studies the classification of life, does so in the context of evolutionary relationships. Cladistics, the predominant method used in systematics today, classifies organisms with respect to their phylogenetic relationships—those based on their evolutionary history. Therefore, students who do not understand evolution cannot understand modern methods of classification.

Developmental biology is another example of a biological subdiscipline enriched by an evolutionary perspective. In fact, some embryological phenomena can be understood only in the light of evolutionary history. For example, why terrestrial salamanders go through a larval stage with gills and fins that are never used is a question answered by evolution.[6] During evolution, as new species (e.g., terrestrial salamanders) evolve from ancestral forms (e.g., aquatic ancestors), their new developmental instructions are often added to developmental instructions already in place. Thus, patterns of development in groups of organisms were built over the evolutionary history of those groups, thus retaining ancestral instructions. This process results in the embryonic stages of particular vertebrates reflecting the embryonic stages of those vertebrates' ancestors.

The study of animal behavior is enriched by an evolutionary perspective as well. Behavioral traits also evolve, and like morphological traits they are often most similar among closely related species. Phylogenetic studies of behavior have provided examples of how complex behaviors such as the courtship displays of some birds have evolved from simpler ancestral behaviors. Likewise, the study of human behavior can be enhanced by an evolutionary perspective. Evolutionary psychologists seek to uncover evolutionary reasons for many human behaviors, searching through our ancestral programming to determine how natural selection has resulted in a species that behaves as it does.

There are many sciences with significant historical aspects, such as anthropology, astronomy, geology, and paleontology. Geology, for example, is the study of the history of the earth, especially as recorded in the rocks. Paleontology is the study of fossils. Inherent in the work of the geologist and the paleontologist are questions about the relationships of modern animals and plants to ancestral forms, and about the chronology of the history of the earth. Evolution provides the framework within which these questions can be answered.

What do science and education societies say about the study of evolution?

Instructors often look to scientific societies for answers to many questions regarding their teaching. There is one aspect of teaching on which the scientific societies agree and are emphatic. Evolution is key to scientific study, and should be taught in the science classroom. The National Research Council, part of the National Academy of Science, identified evolution as a major unifying idea in science that transcends disciplinary boundaries. Its publication *National Science Education Standards* lists biological evolution as one of the six content areas in the life sciences that are important for all high school students to study.[7] Likewise, the American Association for the Advancement of Science identified the evolution of life as one of six major areas of study in the life sciences in its publication *Benchmarks for Scientific Literacy*.[8] The National Science Teachers Association, the largest organization in the world committed to promoting excellence and innovation in science teaching and learning, published a position statement on the teaching of evolution in 1997, which states that "evolution is a major unifying concept of science and should be included as part of K–College science frameworks and curricula."[9] The National Association of Biology Teachers, a leading organization in life science education, also issued a position statement on the teaching of evolution in 1997, which states that evolution has a "central, unifying role . . . in nature, and therefore in biology. Teaching biology in an effective and scientifically honest manner requires classroom discussions and laboratory experiences on evolution."[10] Evolution has been identified as the unifying theme of biology by almost all science organizations that focus on the biological sciences.[11]

So why should students learn evolution? Eliminating evolution from the education of students removes the context and unifying theory that underpins and permeates the biological sciences. Students thus learn disparate facts in the science classroom without the thread that ties them together, and they miss the answers to its underlying *why* questions. Without an understanding of evolution, they cannot understand processes based on this science, such as insect resistance to pesticides and microbial resistance to antibiotics. Students will not come to understand evolutionary connections to other scientific fields, nor will they fully understand the world of which we are a part. Evolution is, in fact, one of the most important concepts in attaining scientific literacy.

6

Questions and Answers:
Science Education

"It seems safe to predict that America will continue
to witness spirited, indeed acrimonious, debates
over the scientific, theological, and political consequences
of evolution for the foreseeable future."[1]

—Ronald Numbers, Historian

The next three chapters discuss typical questions that students, parents, and others ask of science instructors who teach evolution. These are questions we hear directly from students and that instructors report hearing most often; they are illustrative for understanding the mindset of the questioner. Because we contend that it is important for instructors to understand why their students ask the questions they do, we generally provide an explanation of the potential motivation behind the question and sometimes what the questioner is really asking. The three chapters are based on questions and answers related to science education (Chapter 6); religion (Chapter 7); and general education (Chapter 8).[2]

"What do you mean by evolution?"

Very often creationists asking this question (whether in or out of the class-room milieu) are not seeking a technical answer about hereditary characteristics of groups of organisms, descent with modification of different lineages from common ancestors, and so forth. (See the working definition of evolution in the introduction, page 10.) They are usually trying to discern whether the instructor means that the great variety of organisms living today descended from a common ancestor (or, at most, from a few original forms of life).

There are numerous motivations for this question. One is engendered by some forms of progressive theology asserting that different types of organisms evolved over time, but that the gaps in the fossil record indicate where God intervened in the ongoing "natural" process of evolution. The gaps often indicate places where God intervened supernaturally to create new types of organisms that then evolved as the fossil record indicates.

Another common motivation for this question stems from literalist creationists who want to make sure that an instructor is not speaking about horizontal evolution. Many literalists have learned from their creationist leaders that horizontal evolution is totally acceptable within their tenets of creationism. In horizontal evolution, organisms can vary within their supernaturally created "kinds"; for example, dogs may have changed into the various breeds in existence today. However, cumulative changes could never take place that result in descendants beyond (roughly) the phylogenic level of genus or family. Such resultant large change constitutes what literalists refer to as "vertical evolution," the type of evolution that they contend never occurred. (See Chapter 4, page 88). Even the theist creationists often ask the science instructor to explain whether the evolution being taught posits randomness, and if so, to what extent (See Chapter 2).

People who hold these types of doctrines often want to know whether the instructor's meaning of evolution is synonymous with, or at least compatible with, their meaning of evolution. They are motivated to inquire in an attempt to determine whether the science instructor "knows the truth" about what actually occurred or whether the instructor is misguided and attempting to spread the false teachings of evolution to students. In essence, this question is a type of litmus test that is administered to determine whether the instructor is "knowledgeable."

"Is it true that evolution is not based on evidence?"

This question is frequently rooted in the creationist belief that it is *not* evidence that leads people to conclude that evolution is the most accurate scientific explanation of how organisms got to be the way they are. Rather, they believe that one's philosophy about the world—their worldview—underpins this determination. Many creationists who ask this question believe that the evidence for evolution is (at best) shaky. They believe that the scientific community concludes (in spite of the shaky evidence) that evolution occurred because science's underlying philosophy does not consider supernatural causes. Many creationists believe that nearly the entire scientific community is somehow blind to their own current research and to the research produced by thousands of scientists around the world over past decades. They contend that not only are scientists blind to the results of their own work, but also science instructors are incapable of or unwilling to realize this. Moreover, the contention goes further: that science instructors are not pointing out to students and to other interested parties that the scientific community is incorrect.

Professional creationists usually argue that science instructors and scientists have been so indoctrinated—so brainwashed—with naturalistic philosophy in their high school and university courses that they cannot evaluate intelligently the data for evolution as (apparently) creationists can. If scientists and science instructors had not been so philosophically brainwashed, creationists contend, they would be considering and presenting the case for supernatural causality in the science classroom—namely that a supreme being or intelligent extraterrestrial imposed some form of special creation.

Creationists convinced of this sort of rationale are often surprised to learn that there are a great number of church leaders and theologians who attended science classes at secular secondary schools and colleges and who find the evidence for evolution compelling. These individuals are living proof that, contrary to what many popular creationist leaders claim, one can find the evidence for evolution to be convincing yet still believe devoutly in the Bible and be considered upstanding Christians by their families and other members of their congregations.[3]

The converse idea is even more devastating to the creationist claim that science instructors and scientists come to accept evolution simply because they have been brainwashed by their secular education (particularly their secular education in science). There are many persons who accept evolution and who have been raised in Christian fundamentalist literalist homes and who have attended Christian fundamentalist literalist churches, Christian secondary schools, and Christian universities. Even many persons who were raised in homes in which the Bible was read and studied weekly (if not daily) and the Genesis account of creation was taken literally have been compelled, after examining the evidence for evolution and learning how science functions, to agree that the data show that evolution has occurred (and is occurring). These individuals are clearly not, by creationists' own estimations, the products of some humanistic educational system that supposedly brainwashes and indoctrinates its students with the metaphysics of naturalism. On the contrary, many Christian creationists would contend that these individuals were brought up in the "best" way, with a good fundamentalist literalist Christian education. Of course, most of these individuals continue their belief in the Bible and continue to be practicing Christians, but they now believe that God used evolution in His master plan. In the estimation of literalists, these Christians, who once believed in the literal translation of Genesis, are now off the mark. Literalists would now consider it appropriate to attempt to bring their brothers and sisters who have gone astray back to the truth of the literal interpretation of Genesis and a rejection of evolution education.

A small minority of people raised as literalists in home, church, and school find the evidence for evolution undeniable and yet hold onto their prior belief that humans did not evolve, thus maintaining a form of cognitive dissonance. These individuals generally attempt to justify this seemingly indefensible position by stating a rationale something like "God's ways can be far too complex for human minds to understand." Such a rationale is a typical fallback position within many parts of Christendom, particularly among many creationists. The essential meaning of this and similar statements is that the biblical position cannot be wrong. When overwhelming evidence could lead one to a conclusion that seemingly contradicts scripture, the conclusion must be faulty due to human ignorance of God's ways.

Without considering these underlying worldview issues, the direct answer to the question "Is it true that evolution is not based on evidence?" is no, it is not true. Evolution is based on a vast amount of empirical evidence. The easiest way to illustrate that there is a staggering amount of evidence supporting evolution is by having the questioning student conduct a search of scientific journal articles, books, and research conferences on the subject. S/he can then total the number of "hits." Another interesting exercise is to have the creationist student conduct a historical study of why the scientific community of the early 1800s considered creationism compelling, and why the scientific community of today finds evolution compelling. To accomplish this task, the student can examine discussions of the evidence for evolution reported in scientific publications to date. Usually by the end of such exercises, the creationist still may believe that there are underlying causal philosophical differences between creationism and modern science, but the student usually realizes (at the very least) that a stunning amount of empirical evidence exists for evolution. (These suggestions assume that the creationist with whom the instructor is dealing is not only a student of his or hers but also is willing to take on this "extra credit" work.)

"How can you teach something that no one can see?"

The person asking this question generally is not satisfied with examples such as the evolution of pathogenic microbes. Rather, the person is usually referring to large-scale changes such as those of invertebrates to vertebrates. Because they know that science cannot directly observe large-scale evolutionary changes occurring within human history, they think that evolution should not be taught as a confirmed theory but only as mere speculation. Creationists prefer to define "science" as "knowledge," and they argue that one cannot know something scientifically unless one has seen it. They have the impression that factual theories (not a contradiction in terms—see Chapter 4) in science should be observable, and if they are not, then they cannot be considered either a theory or a fact of science. Since no one has ever seen large-scale evolution take place, how can anyone possibly teach evolution as scientific fact?

Like many of the leading creationists, the student who asks this question also may argue for "creation science" to be included in the science curriculum. What we believe the leading creationists do understand (but what the creationist student almost always does not and finds surprising when revealed) is the following creationist contradiction. Leading creationists state that both creation and evolution are not scientific because we cannot observe them occurring. Yet they argue that creation should be taught in the science curriculum. Two leading creationists write the following:

> If we really *could* see things evolving from, say, a given species into another species of greater organized complexity, we would all have to believe in evolution. We could verify it by observing it—*that* would be real science! But since we *cannot* see it functioning, it is not any more "scientific" than creation.[4]

So apparently, leading creationists openly admit that creationism is not "real science." We would certainly agree that creationism is not real science and that is why we argue that it should not be included as science in science curricula. However, we would disagree with their contention that direct observation is a criterion for factuality in science. The creationists have been so successful in popularizing this criterion to the general public that some legislators appear to have been influenced. The following message from the Alabama State Board of Education has been placed in Alabama public school textbooks:

> This textbook discusses evolution, a controversial theory some scientists present as a scientific explanation for the origin of living things, such as plants, animals and humans. No one was present when life first appeared on earth. Therefore, any statement about life's origins should be considered as theory, not fact.[5]

Many students, creationists or not, characterize science as requiring direct observation and prediction (in the sense of future observation). However, science does not require this.[6] The historical sciences (e.g., astronomy, geology, paleontology)—and even biblical archeology—often use a different type of inference, namely the scientific principles of post-diction and retrodiction. In these processes, scientists determine how well a theory explains the data pattern they are directly observing.

The student might be encouraged to consider that no scientist has ever directly observed a dinosaur—just fossils of "supposed" dinosaurs. We infer that the fossils we examine today are from organisms that lived in the past. Even most creationists agree that fossils are evidence of past organisms living on the planet, including all the various dinosaurs. So even though scientists did not observe living dinosaurs, their existence is nonetheless considered a scientific fact. Likewise, humans did not directly observe the evolution of the dinosaurs, but their evolution is nonetheless considered to be scientific fact. Obviously there are thousands of examples of matter and processes that have not been directly observed but that nonetheless are considered scientifically factual by both creationists and evolutionists, often including those in biblical archeology (and all of human history with no living eyewitnesses). However, when it comes to the process of evolution, creationists say that direct observation is a criterion for "real science." The student asking the question "How can you teach something that no one can see?" needs to understand that historical science research is not conducted by using direct observation of occurrences in the past. Expecting direct observation to be required when studying ancient macroevolution is unreasonable and an exception to normal procedures used in historical science research.

"If organisms evolved, then why do they look so well designed?"

This question is one of the most common questions asked by creationist students. To the person asking, the organisms on our planet appear to operate extremely well. They operate so well, in fact, that it seems absurd to them that the science instructor would even put forth evolution to explain what appears to be so clearly designed (i.e., proximately designed).

Often, the first step for science instructors is to explain how something familiar could *appear* to have been designed for the current use, but, in fact, may not have been. The following story well illustrates this fact: Long ago we observed an advanced first aid class for health-related professionals in which the instructor was explaining the action to take when a person has a blocked airway. After exhausting the list of multiple ways to clear the airway, the instructor taught the technique of last resort—the tracheotomy. Because these professionals worked in the streets and could find themselves without even the most basic of first aid kits, the instructor illustrated how a Bic pen might be used—if no other option remained—to establish an airway (when stripped of its ink cartridge, cap, and plug). It struck us how an extraterrestrial visiting our planet, who possessed similar intelligence and experience but not an identical industrialized history, might think about the Bic pen after attending this first aid class. The extraterrestrial could conclude that Bic had *designed* their apparatus to work as an emergency human airway. Furthermore, the extraterrestrial, upon examining commercial buildings and private homes throughout North America, might believe that company designers gave the Bic product the secondary function of a writing implement and maintained its low price to encourage the pens' omnipresence for use in life or death emergencies. Put simply, the extraterrestrial might reason that the pen's use as an emergency airway was the primary reason for its design because it worked so well for this purpose.

Distinguishing present utility from reasons for origin is counterintuitive to many students. Therefore, using multiple examples, employing various methods, and taking sufficient time is often necessary for student understanding. (For more advanced students, see Gould & Lewontin for examples from architecture to anthropology.[7])

The second step science instructors might take to answer the above question counters proximate intelligent design in biology with evolution via natural selection. Most students who are challenging evolution with an intelligent design question/argument, are generally unaware of counterexamples. Biologist Kenneth Miller puts forth the following:

> Careful studies of the mammalian fossil record show that the average length of time a species survives after its first appearance is around 2 million years. Two million years of existence, and then extinction. The story is similar for insects (average species duration: 3.6 million years) and for marine invertebrates (average duration: 3.4 million years). In simple terms, this designer just can't get it right the first time. Nothing he designs is able to make it over the long term.[8]

Miller is a Catholic who accepts evolution and holds his religious faith. He writes to illustrate how a scientific explanation of design to counter evolution would be flawed. Instead of past species possessing such good design that would allow them to survive at least until humans arrived on the scene, most became extinct. Not only did they become extinct, but also their extinction occurred rapidly (relative to how long life has existed on earth). So approximately everything that the Intelligent Designer designed quickly died out—the design was insufficient for survival.[9]

If appropriate in their particular situations, science instructors may ask the student to offer a *scientific* explanation for this phenomenon—that is, to "square" the extinction data with the idea of intelligent design. The key to answering this question is, of course, to open the eyes of students so they can see that explanations involving intelligent design are scientifically far less compelling than the evolution they wish to refute. When intelligent design students *do* offer explanations for the extinctions, they will most likely be theological ones—"God wanted it that way" or "God's ways are not our ways." When students rely on theology to posit a scientific explanation and this reliance is pointed out to them, they often begin to see the inconsistencies in their anti-evolution argument from design. Even nonreligious persons who want intelligent design in the science curricula often begin to realize that they must make up stories about the desires of the extraterrestrial intelligence to explain extinction phenomena, such as maybe the designer wanted an organism to die out—become extinct.[10]

Some creationist students who can offer more advanced argumentation concerning intelligent design may posit that there is an analogy between NASA's search for extraterrestrial intelligence (SETI) and creationists looking at biological structures (e.g., DNA) for signs of extraterrestrial intelligence. They may claim that because SETI is seeking to detect complex information attributable to intelligent design, biologists should concede that DNA, for example, is complex information that should be likewise attributable to intelligent design. On the contrary, nature is full of structures that contain extremely complex patterns that science attributes to natural processes, and DNA is one of them. Students need to understand that chance can produce complex information, and that structures such as DNA and the information it contains can be maintained and channeled by natural selection. Furthermore, science considers evolution to be a natural process that may occur regularly on other hospitable planets. If the extraterrestrials we are searching for exist, then they would be products of natural processes—evolution. In short, it is precisely because of the natural process of evolution that some scientists think extraterrestrial life might exist. What SETI is attempting to detect is complex information sent from a conscious intelligence in a form that scientists can currently understand. DNA is clearly not this type of information/language.[11]

"Why can't intelligent design theory be included in science curricula?"

Creationists want intelligent design to be included in public school science curricula as an alternative to evolution by natural selection, rather than advocating the removal of evolution from the curriculum altogether. What students should know is that science is a human construct, unlike nature (even though we "access" it via our senses and mind). This human construct is an aggregate of those who do the constructing—the scientific community. Anyone can observe the current construct of science by examining scientific journals, which report what the scientific community is doing—science. When we teach science to students, we teach them what the scientific community does. To teach something different is akin to saying nonsensically to our students: "This is science even though the scientific community does not do it."

Does the scientific community include intelligent design in scientific explanations? George Gilchrist, a University of Washington evolutionary biologist, conducted a search of the primary science literature—approximately 6,000 journals in the life sciences—for "intelligent AND design." His results:

> This search of several hundred thousand scientific reports published over several years failed to discover a single instance of biological research using intelligent design theory to explain life's diversity.[12]

(During this search, Gilchrist also looked for scientific research articles containing the words "creation science"—he found none.) Barbara Forrest, a philosophy professor at Southeastern Louisiana University, replicated and expanded Gilchrist's intelligent design survey, which yielded the same results—intelligent design proponents fail to publish in the scientific literature.[13] Gilchrist asks an extremely pertinent rhetorical question: "Why should we reserve a place in the science curriculum for science that apparently does not exist?" The answer is that we obviously should not. This is another point that students should come to understand.

The intelligent design creationists generally respond to this line of reasoning by saying that the scientific community is either suffering from materialist blinders (and therefore can't see intelligent design when it is present) or is conspiring against admitting to intelligent design (or some combination of the two). Our response has always been that if what you believe is really going on in the scientific community, then that is where your battle lies—in the scientific community. The creationists must first change the construct of the scientific community; then science instructors will teach intelligent design because it is part of the construct. Until that day, instructors cannot honestly teach it as science. Since the creationists have continually failed in their attempts to muster enough evidence or methodological justification to have intelligent design considered appropriate or useful by the scientific community, they are attempting to inject it into public school science curricula as science via the back door.

"Because scientists don't know every detail of how evolution occurs, shouldn't they at least consider supernatural causes as scientific explanations and teach such possibilities in the science classroom?"

No. Just because we currently may not have a scientific explanation for every aspect of every phenomenon does not therefore require that we invoke supernatural causes and teach them as science. Scientific explanations are different from religious explanations, and many highly religious scientists have no problems conducting their scientific research while maintaining their religiosity. Even some renowned scientists who are Christians believe for religious reasons that God may be involved with

evolution, either by designing the initial laws or systems that make evolution possible or by intervening occasionally in the evolutionary process.
But these scientists make a distinction between scientific explanations and
religious explanations. For example, Owen Gingerich, Professor of
Astronomy and History of Science at Harvard, contends the following
with respect to evolution, science, and his Christian faith:

> As a Christian theist, I believe that this is part of God's design. Whether
> God designed the universe at the outset so that the appropriate mecha-
> nisms could arise in the course of time, or whether God gives an occa-
> sional timely input is something that science, by its very nature, will
> probably never be able to fathom. But as a scientist, I accept evolution as
> the appropriate explanatory structure to guide research into the origins
> and affinities of the kingdoms of living organisms.[14]

Gingerich, even though a Christian who sometimes criticizes the suf-
ficiency of evolutionary explanations, has no trouble distinguishing
between religious and scientific explanations while believing that both are
useful. He understands that while anomalies do exist in science, that
invoking supernatural explanations for those anomalies is inappropriate
and scientifically profitless. Supernatural explanations are scientifically
profitless because they provide causal relationships (supernatural relationships) that science cannot examine. One cannot test for supernatural relationships with naturalistic equipment and methods. If science did
entertain supernatural causes, then any phenomenon not yet explained by
empirical evidence and arguments leading to credible naturalistic causal
explanations could be explained by invoking the supernatural. However,
supernatural models of explanation cannot be tested. And even if the phenomenon being studied had sufficient empirical evidence and arguments
to compel all but one person of its naturalistic relationship (e.g., one
magnet causing another magnet to move), that one person could still put
forth a supernatural explanation. Nevertheless, the relationship between
the one magnet and another magnet can be examined while the relationship between a supernatural power and the magnet cannot.

Students should be shown that in their scientific writings, evolutionary
biologists discuss proximate nonsupernatural causes. Students should be

given a chance to recognize that while these scientists agree that evolution occurred, they still debate the extent to which various mechanisms cause evolution. Even when scientists debate a main cause (not ultimate cause) of evolution, the discourse remains proximate. For example, evolutionary geneticist Graham Bell writes that mutation and sampling error should be considered less of a causal explanation for evolution, while the mechanism of selection actually provides the primary cause of evolutionary change:

> Evolution is caused by selection. Mutation and sampling error are important processes that act directly to reduce the effectiveness of selection and indirectly to fuel selection. They are not themselves theories of evolution capable of explaining the organization and history of living organisms.[15]

When students begin to see the actual discourse that scientists employ when explaining or debating the main causes of evolution, the inappropriateness of inserting supernaturalism into such explanatory discourse may become more obvious.

Just because most scientists understand that supernatural causes cannot be entertained in science does not stop some creationists from arguing and popularizing the opposite. For example, intelligent design advocate Michael Behe's views were discussed in an article in *The Chronicle of Higher Education* entitled "A Biochemist Urges Darwinists to Acknowledge the Role Played by an 'Intelligent Designer.' "[16] Our response published in the same journal might be illustrative (edited for inclusion here):

> It is certainly nothing new that some scientists, like Michael Behe, argue that "science has made a fundamental mistake by trying to explain the world exclusively in physical and material terms." However, purported novelty aside, one of the many problems with introducing supernatural causality into science is the problem caused by following it through to its logical applications and conclusions. For example, the scientific investigation . . . by the Central Intelligence Agency, the Federal Bureau of Investigation, and the Federal Aviation Administration to determine the cause of the crash of TWA Flight 800 would have had four, not three, hypotheses: 1) mechanical failure; 2) on-board bomb explosion; 3) missile impact; and 4) supernatural intervention. Furthermore, forensic scientists testifying in the O. J. Simpson trial would have had to consider

and communicate to the jury the possibility of supernatural intervention as a cause of the homicides.

Sounds irrational? Of course it does. Just because science may not currently possess a natural explanation of a phenomenon that is satisfying to some people, it does not follow that supernatural causes must be considered. If science currently knew nothing about the germ theory of disease, would we be content with acknowledging an "intelligent designer" of cholera, or would we have science press forward to try to find a natural cause? As for us, please pass the tetracycline.[17]

Most creationist students seem to understand intuitively that scientific investigations concerning very serious matters such as airplane crashes, homicides, and disease outbreaks should not involve the construction of hypotheses about supernatural causes. Therefore, it can be quite useful to have small groups of students role-play as forensic investigators and other types of scientists before the word *evolution* is introduced in a course. Due to the nature of some courses, this means conducting the role-play during the first few meetings, which may be beneficial in introducing various methods used in science. In their roles as scientists, the students are confronted with the task of constructing defensible explanations to a phenomenon such as those mentioned previously. The "scientists" offer their explanations orally to their "superiors or supervisors," or can write their explanations for "journal submission." Nearly every student will respond with explanations that do not involve the supernatural. There are, of course, the occasional few students who use *X-Files* kinds of explanations in an attempt to be comical, but, after encouragement to be serious, these students typically give natural explanations as do the other students. The class is then asked why they did not offer, in their role as scientists, supernatural explanations? Given sufficient time, the students tend to answer the question by constructing lines separating what is appropriate to be offered as scientific explanations and religious explanations. After some discussion, the class usually comes to an agreement about what constitutes the demarcation between natural and supernatural explanations and a resulting rubric is produced and recorded. Throughout the rest of the course, whenever students think that a supernatural explanation (such as creationism) should be dealt with in science, the rubric reappears and is applied to their concern. The students generally then recognize the discrepancy and, over time, come to know (sometimes reluctantly) why the overwhelming majority of scientists exclude creationism from science.

"Why is evolution considered a scientific fact?"

Two misconceptions generally underpin this question: (1) evolution cannot be a scientific fact because it is a theory, and (2) evolution certainly can't have the confident status of scientific fact. Creationists usually do not understand that evolution is considered scientifically factual for the same reasons as is the past existence of dinosaurs. Both are upheld by such an overwhelming body of evidence that the vast majority of the scientific community is compelled to conclude that dinosaurs existed and that we evolved. Contributing to these misconceptions is the fact that creationists, as well as a large percentage of noncreationist students, continue to use the word *theory* in its colloquial sense even when speaking about scientific theories. They incorrectly believe that even long established scientific theories are just educated guesses with little or no evidence to uphold them. They are "just theories." Ernst Mayr, Professor Emeritus of Zoology at Harvard University, defines facts in relation to theories:

> Facts . . . may be defined as empirical propositions (theories) that have been repeatedly confirmed and never refuted.[18]

Many science instructors find it convenient to explain to students that, in general, scientific laws are descriptions of phenomena and scientific theories are explanations of phenomena. Both laws and theories have equal potential for factuality. So prior to attempting to explain the evidence for evolution to students, it is useful to define facts, theories, laws, hypotheses, and concepts, particularly in relation to biology. Moreover, many concepts in biology are as worthy in status and usefulness as are laws in the physical sciences. Mayr states that "Today, concepts such as competition, common descent, territory, and altruism are as significant in biology as laws and discoveries are in the physical sciences."[19]

After students understand that evolution can be considered a scientific fact (in principle), the next step is teaching the evidence that compelled biologists over the years to accept it as fact. Mayr recounts:

In 1859 Darwin's ideas about the inconstancy of species and common descent were considered to be theories. The amount of evidence in favor of these "theories" and the absence of any counterevidence has, since then, led biologists to accept these theories as facts.[20]

(Scientific theory, law, and fact are also discussed in Chapter 4, page 87.)

"Why can't you prove evolution to me?"

Like much of the general public, many creationists somehow think that the research concerning evolution lacks the scientific rigor to be compelling to the point of "proof." The creationist is not using this term in the mathematical sense of deduction in a proof, but rather in the colloquial sense of "it must convince me." Using this meaning of proof, creationists do not understand how science instructors can teach that biological evolution is confirmed while creationists are not compelled to agree.

This nonunderstanding has more to do with the definition of the word *proof* and less to do with evidence and argument. Scientists, as well as creationists, usually are not referring to the deductive proofs found in the disciplines of mathematics and logic, but, unlike creationists, scientists use the word in the sense of evidence or confirmation. When scientists say that they have proof of X occurring, they mean (more precisely) that they have such compelling evidence that X's occurrence would be considered confirmed in the scientific community. This does not mean that 100% of scientists will agree that the occurrence of X has been confirmed—after all, there are a few scientists alive today who find plate tectonics, heliocentricity, and the past existence of dinosaurs unconfirmed. (They believe that aliens or supreme beings deposited dinosaur fossils.) Many scientists do not use the word *proof* at all. They prefer to say that they uphold hypotheses and theories with evidence.

Many creationists think that when something is confirmed in the scientific community, the reasons for its confirmation should be blatantly obvious to someone outside the community. They tend not to understand that sometimes it takes significant time and training to understand how all the evidence contributes to the confirmation of certain concepts in science. Understanding the underlying evidence for certain phenomena in astrophysics or genetics, for example, may take students years of their college educations to begin to comprehend.

While some areas of evolutionary research may be in this category, many areas are not and can be understood by high school students and beginning undergraduates. Students must understand that just because someone does not accept the compelling evidence for evolution, it does not follow that the evidence is not compelling for the scientific community. In addition, students should learn that building coherent patterns of explanation is, in itself, a major portion of the endeavors of the scientific community. Moreover, evolution is certainly the reigning (or only) acceptable scientific explanation for the diversity of life. Part of the job of science instruction is to teach students why the scientific community finds the explanatory patterns of evolution acceptable and the evidence for evolution compelling to the point of scientific confirmation, and—in the creationist colloquial sense—"proved."

"What good is a partial eye, wing, or other structure?"

Many students—not just creationist students—ask the common-sense question: "Before current eyes and wings evolved, what good would a partial eye or wing be to an organism?" The wing and eye are overwhelmingly the most asked about structures from students and professional creationists—another correlation between the protests of professional creationists and the questions science students commonly ask about evolution. Professional creationists attempt to convince students that any "partially evolved" structures (creationists' understanding), which may have existed prior to current structures in today's living organisms, would be detrimental and therefore undermine evolutionary theory. Henry and John Morris write in one of their popular books:

> A true transitional structure would be, say, a "sceather"—that is, a half-scale, half-feather—or a "ling"—half-leg, half-wing—or, perhaps a half-evolved heart or liver or eye. Such transitional structures, however, would not survive in any struggle for existence.[21]

The misconception underpinning this question is that only the eye and wing as they currently exist are usable. The short and simple reply is that eyes "partially evolved" in comparison to their current state could probably see partially as well—much better than no eye at all. Science instructors might find it useful to have students conduct a literature search to find out whether intermediate structures would be useful and adaptive for such organisms. Students will find the scientific literature rife with examples that partial wings allowed organisms to drift, glide, float, and so forth. In addition, students will find that these primitive wings might have had uses other than flight; they could have been used as heat exchangers, for example, with feathers acting as insulators. Most students are simply not aware of these facts when they ask questions about partially evolved structures. When given the opportunity to learn about the arguments concerning wing evolution (insulating feathers to aerodynamic wings) and eye evolution (light-sensitive spot to image-focusing eyeball), most of the students then understand the logic of these evolutionary explanations.

However, it is significantly more difficult for many students to understand that scientists do not have conclusive adaptive explanations (via selective force) for all structures currently seen in organisms. Rather, their explanations involve more complex matters of organisms' evolution as a whole, such as morphological designs constrained by phyletic heritage, developmental pathways, and general architecture. (For further information see Gould & Lewontin [1979]; Pinker [1997b]).[22] Naturally, these are important matters for students to learn, but the subject, nature, and level of the course must be taken into consideration for appropriateness. Usually when students bring up the question of the usefulness of partially evolved structures, they are not seeking answers concerning the more sophisticated matters of analyzing organisms as integrated wholes but are rather simply interested in what general usefulness a partial eye or wing could be to an organism.

"Isn't evolution a theory in crisis?"

Creationists who have read creationist literature and/or attended creationist lectures generally come away with the impression that soon evolution will be disconfirmed within the scientific community. The reason creationists develop this impression is that the creationist movement (particularly literalist and progressive) favors pointing out disagreements among scientists in the scientific literature and elsewhere. Creationist authors often take brief quotes (a sentence or two) from numerous sources, that, when read out of the context of the scientists' articles and juxtaposed with creationist pre- and post-quote commentary, appears as if the scientists are arguing against the occurrence of evolution. (See Chapter 4, pages 90–94.) Here's the way many creationist students launch the attack:

> The evolution you are teaching in the classroom is mostly false. Contrary to what you are telling us, natural selection is not involved in evolution, there are no transitional forms, and there is no evidence for evolution having occurred in the fossil record. Harvard University paleontologist Stephen Jay Gould writes:

> "Paleontologists have paid an exorbitant price for Darwin's argument. We fancy ourselves as the only true students of life's history, yet to preserve our favored account of evolution by natural selection we view our data as so bad that we never see the very process we profess to study."[23]

> Professor Gould is one of the world's leading paleontologists and tells us about the dearth of evidence for evolution. So there's no evidence for evolution.

Those who know of Stephen Jay Gould's voluminous work concerning evolution and those who have read the full article from which this one sentence was extracted ("Evolution's Erratic Pace") know that Gould is certainly not casting doubt on whether there is compelling confirmatory evidence for evolution's occurrence. Instead, he is arguing that the paleontological data for an always smooth and gradual evolution is lacking, and that the data often better support evolutionary lineages characterized by brief episodes of rapid speciation separated by long intervals of stasis (i.e., Eldredge and Gould's theory of punctuated equilibrium). But when a creationist presents only an out-of-context quote couched among creationist commentary, it is understandable that students might be confused as to whether the scientist was questioning the occurrence of evolution itself. When people are subjected to a barrage of such out-of-context quotes, there certainly could be a tendency for them to believe that evolution is a theory in crisis within the scientific community.

Students holding such perceptions should be told that scientists are not arguing about whether evolution occurred but rather are debating various hypotheses concerning specific aspects of the mechanisms or pathways of evolution. Science has no disagreement about the fact that evolution occurred.[24] Additionally, it is useful to show students that it is not only evolutionary scientists who argue about varying hypotheses, but scientists in all fields who argue on a regular basis in scientific journals and elsewhere over varying hypotheses related to numerous confirmed concepts. Of course, leading creationists do not contend that these other confirmed scientific concepts are in crisis (those with which they agree)—just evolution, and any science that corroborates it.[25]

Appendix A contains a list of selected North American scientific organizations that have attested that evolution is scientific and that support its teaching in science courses. None of these organizations supports creationism being taught in public school science courses.

"Do you evolutionists want students to think evolution is scientifically correct just because most scientists do?"

Many creationists will say that science instructors argue from authority by talking about what scientific organizations conclude. The person asking this question often assumes that science instructors who teach evolution as fact are attempting to get students to think they are correct by an argument of authority alone. They frequently think this way because they know that the overwhelming majority of scientists find evolution to be the *only* scientific explanation concerning the diversity of life. Moreover, usually in their educational backgrounds, instructors or friends have hinted that they must be irrational for rejecting evolution because (this is the offending portion of the causal argument) the majority of the scientific community conclude it to be accurate. Creationists do not like what they perceive to be merely a democratic argument as to why evolution is taught. Often, noncreationists also take offense with such an educational rationale.

Creationists say they want to base conclusions on evidence. What they tend to misunderstand is that evolution is *not* a theory in crisis precisely *because of* the evidence. It is evidence, not arguments from authority, that underpins the acceptance of evolutionary theory by the scientific community. If this were not the case, then why would evolutionary scientists be studying, examining the research literature, running experiments, and collecting data? Creationists want to believe that all the leading scientific organizations are somehow in a conspiracy to support evolution despite no evidence, or are deceived into believing that evolution is on sound footing. What creationists will not accept is the idea that leading scientific organizations know evolution is scientifically confirmed because of the incontrovertible evidence, and that is why the scientific community considers evolution to be scientifically factual.

While it is certainly proper for science instructors to indicate to students that most scientists find the evidence for evolution compelling, they should emphasize far more the reasons *why* scientists are compelled to conclude that evolution occurred and is occurring. Science instructors can show students that evolution is so scientifically credible that it is virtually impossible to find an article in a recognized scientific research journal arguing against evolution. They might also point out to students that while curriculum designers may make decisions on what to include in science curricula based on what the majority of the scientific community finds to be factual and important (generally including broad and overarching concepts), science instructors do not (or should not) teach students to think that evolution is factual just because others do. They should teach *why* the others do—the scientific justification for the conclusions.

Finally, a desired emphasis throughout science instruction is fostering the development of process and reasoning skills that scientists use in their work rather than asking students to learn *just* scientists' conclusions. Having students understand the processes by which conclusions are drawn is more important than having them "get the right answer" by mere memorization or illogical reasoning. The spirit of this view is held by most science educators and is also held by some historical figures in evolution: T. H. Huxley wrote, "Irrationally held truths may be more harmful than reasoned errors."[26]

"Why shouldn't scientists invoke the supernatural for creation of first life?"

Quite often creationist students and a large section of the general public come to believe that the scientific community is entirely dumbfounded when it comes to explaining how inanimate matter became living via natural laws and, therefore, scientists should invoke supernatural causes in science. Creationist students typically come by this belief through professional creationist ideas that are passed, both verbally and in print, within their nonschool environments (or their school environments if they attended a creationist school). The picture that is painted has scientists reluctantly admitting they lack any possible naturalistic explanation for the origin of life but still think it occurred because the alternative—creationism—is unacceptable due to some evolutionary ideology and/or their lack of belief in God.

There are at least two matters that should be addressed concerning such misconceptions. First, scientists *do* have reasonable scientific hypotheses for the origin of life (see Origin of Life, Chapter 4, page 98). Second, many scientists *do* believe in a personal God, yet seem to have no trouble conducting naturalistically based research. (Roughly 40% of scientists are in this category. See The Unbelievable Battle, Chapter 1, page 31).

Helping students understand how life might have emerged via naturalistic causes is a very important part of the reply to this question. Despite such attempts, many students hold to their misconception that science has no hypotheses about the origin of life, basing this erroneous conclusion on the authority or stature of prominent scientists whose comments have been taken out of context. Additionally, some creationist students compare what they believe God says and what their science instructors are suggesting, typically confusing the separate questions and domains of science and religion. In such a case there is often no easy quick solution (see Category Solution, page 58). Yet, in many cases, creationist students *do* want to discuss possible proximate naturalistic causes and have had some exposure to professional creationist literature. In these cases, science instructors are being asked to counter authority arguments constructed from out-of-context citations of evolutionary scientists that have been strung together to make an argument against evolution and for supernatural causation. In the case of origin of life, creationists might string together quotes from evolutionary scientists debating mechanisms for the origin of life, resulting in a perception that the scientists are nearly concluding that life could not have originated via natural causes.

Many students who read such literature perceive that these quotes give them an edge in their argument with their science instructor. In many young students' minds, the science instructor's explanation is considered far less credible than those of the evolution experts (whose words have been taken out of context). Although it is best for instructors to encourage students to consider the issues based on the merits of the argument and not on the authority of the advocate, the first step with many creationist students is to have them confront the fact that the evolutionary experts they cite currently conduct research on naturalistic explanations for the origin of life. These scientists have not stopped their work because they think their searches are in vain. Many students need to be confronted with this obvious point and hear that leading scientists *do* find the origin of life scientifically explainable. For example, the following quote from Ernst Mayr is often surprising to many—"In fact, there is no longer any fundamental difficulty in explaining, on the basis of physical and chemical laws, the origin of life from inanimate matter."[27] To most science instructors, Mayr is stating exactly what we would expect him to state. Many creationist students find such a statement to be more challenging to their authority argument than science instructors' best-reasoned explanations.

"What should we do if empirical evidence and materialist philosophy are going in different directions?"

This question comes from anti-evolutionist Phillip Johnson's book *Defeating Darwinism*, a book he wrote for advanced high school and beginning college students. The book is his attempt to show students how to attack the teaching of evolution by attempting to drive a wedge between what he calls "scientific materialism" and facts revealed by scientific investigation. Johnson instructs students and others to:

> Just push this question and refuse to accept the usual evasions as answers. "What should we do if empirical evidence and materialist philosophy are going in different directions?"[28]

It seems somebody is reading Johnson's book because science instructors have reported this question being asked many times by creationist students in their classrooms.

To respond to the students and parents who might use a type of Johnson "wedge" strategy, we provide the "wall" defense. The wall simply represents the separation of one type of knowing from another—those explanations based on natural occurrences and those based on supernatural occurrences—to a level appropriate for late high school and beginning college students (Johnson's readers). Here is an example: Suppose you were on a jury in a trial of a man accused of burglarizing a bank vault. The evidence against him is his fingerprints at the scene and the bank-marked bills found hidden in a closet at his home a block away from the bank. The defendant states to the jury that he has never been to that bank nor does he know how the money got in his closet.[29]

The defendant's attorney contends that his client is innocent but explains to the jury that his client's fingerprints were found at the scene of the crime and that the stolen money was found in his closet. The defense attorney states that because he believes his client is innocent, he asks you to consider that the fingerprints at the scene of the crime were created and that the bank money was transported to the closet, both via some supernatural cause. Naturally, one would be hard pressed to find a juror—creationist or not—who would consider the supernatural explanation to be scientific testimony, even if an "expert" witness gave a supporting opinion. The point here is not whether the jury would find the man guilty or innocent of the crime but rather that the legal profession would consider it unreasonable to entertain supernatural causality in a court of law. After all, one does not hear of "playing the supernatural card" (supernatural strategy) among the legal establishment.

An interesting twist on the above scenario is to suppose that the defendant says he was working 2,000 miles away from where the robbery occurred that day. However, additional evidence exists that he committed the crime. Five witnesses testify that they saw the defendant commit the crime. Nevertheless, in his defense, an office full of his fellow employees, all with good reputations, testify that they saw him in his office throughout the entire day of the offense except for a 10-minute work break when he was out of their sight. Some even testify that they spoke with him in person during that day. Shockingly, the prosecution agrees with the office employees that the defendant spent the entire day working in his office but also contends that during that 10-minute break he went into a closet and was transported 2,000 miles away to the site of the crime via some unknown, unidentifiable supernatural occurrence. He then committed the robbery and afterwards was supernaturally transported back to the closet and returned to work. It is probably difficult for most of us to even imagine such an outrageous suggestion ever coming from a prosecutor in a modern court of law. But if it were to be put forth, it appears that following Johnson's reasoning, we would be asked to seriously consider the supernatural closet transport as science. After all, Johnson encourages students to press the question: "What should we do if empirical evidence and materialist philosophy are going in different directions?" With regard to this scenario, it appears that he is asking, "What should we do if the identity recall of the 5 witnesses and the science of space-time movement of humans are going in different directions?" Our answer is we would *not* entertain the supernatural possibilities as science.

Now let us introduce more empirical evidence into the supernatural closet transport scenario to take it a step further. Collected from the burglary scene were positively identified fingerprints, DNA evidence, and a videotape all attributed to the defendant committing the robbery. (The reason we are adding more empirical evidence into this scenario is to attempt to simulate more closely what the anti-evolutionists contend—that there is significant empirical evidence against evolution.) In addition, all the robbery eyewitnesses were lifetime acquaintances of the defendant. Furthermore, all 5 witnesses not only saw the defendant at the time and place of the crime but also talked to the defendant while he was committing the robbery. As in the previous scenario, the prosecuting attorney puts forth his supernatural closet transport. Even given this large quantity of empirical evidence that the defendant was present at the robbery, would an anti-evolutionist sitting in the jury consider the supernatural closet transport explanation scientific? Could the creationist possibly find the defendant guilty of the robbery even though believing he was 2,000 miles away all day except for ten minutes? Whether in the courtroom, in the laboratory, or in the science classroom, there is a wall separating consideration of natural occurrences from supernatural occurrences. That wall is the one prudent way, and many would say a necessary way, to separate science from non-science. (See Stephen Jay Gould's *Rocks of Ages* for a lengthy discussion on a respectful separation of science and religion.)

7

Questions and Answers: Religion

"A 'Christian evolutionist' . . . is an oxymoron"

—Henry Morris, President Emeritus, ICR[1]

"Didn't Darwin recant on his deathbed?"

It is amazing how widespread the legend is of a Darwinian deathbed recantation. As the story goes, Darwin admitted he was wrong about evolution, became a believer in God, and asked for God's forgiveness. The legend thrives in some evangelical books, tracts, and magazines, and is aired occasionally on television.

There are at least three significant problems with this recantation story. First, Darwin thought that a person could accept both the scientific fact of evolution and still believe in a God. Second, Darwin was never an atheist. Just a few years before his death, Darwin wrote that he had "never been an atheist in the sense of denying the existence of a God."[2] In fact, he tells us that he was a Christian for many years in this statement: "I never gave up Christianity until I was forty years of age."[3] Disavowing Christianity certainly does not mean one necessarily gives up belief in God—although most Christian creationists would contend that without Christianity there is no salvation, only damning judgment by God. So, for these creationists, Darwin's loss of Christian belief is the primary point of concern.[i] Christian creationist Henry Morris helps illustrate that viewpoint when he writes that Darwin "became, to all intents and purposes, an atheist."[4] But that is generally not how historians have viewed Darwin. Thomas Henry Huxley coined the term "agnostic" in 1869,[5] and 10 years later Darwin began using this term to describe himself.[6]

Third, there is no reliable evidence that Darwin ever made a so-called deathbed recantation. In the early 1900s, the story swept through fundamentalist America via articles in Christian magazines and tracts with titles such as "Darwin on His Deathbed," "Darwin's Last Hours," "Darwin, 'The Believer,' " and "Darwin Returned to the Bible." Although the stories had immense publicity, their writers rarely contacted the Darwin family to see if the story was accurate. When contact was made, the story was denied vehemently.[7]

Taking these three points into account, a science instructor would be hard pressed to answer the question of Darwin's deathbed recantation with anything other than a denial. However, since creationist students often consider creationist leaders to be more authoritative then noncreationists, science instructors might consider also citing Henry Morris's opinion about Darwin's supposed renunciation of evolution: "The evidence for this [Darwin's recantation], however, is doubtful."[8]

As far as the practice of modern evolutionary sciences are concerned, it does not matter what Darwin might have said on his deathbed. Students should realize that even if Darwin had stated he was completely wrong about evolution, it would make no difference to today's science. Scientists do not believe evolution occurred because Charles Darwin said it occurred. Evolution is confirmed because of the overwhelming evidence and explanatory power of the theory. The evidence we possess today is far greater in quantity and quality than was available to Darwin or even to the

---

[i] Many Christian fundamentalist denominations have a doctrine of eternal salvation, which means that Christians can never lose their Christianity. Therefore, if Darwin had been a Christian, he would have never lost his salvation and would therefore have entered heaven. However, some of these same denominations would also hold that a true Christian believer never would have disavowed his Christianity in the first place. Therefore, they would contend that Darwin was never a true Christian.

entire scientific community at his time. Because Darwin's deathbed question is irrelevant to the controversy concerning evolution education, questions about the legend really do not seem to matter. But to the student asking the question, it usually does matter, at least at the moment.

"Doesn't evolution necessarily lead to moral decay?"

There is an assumption among most creationists that accepting the occurrence of evolution frees people from moral responsibility. The argument is naive, but still widespread among students and adults alike. The argument proceeds: If people thought that they descended from animals, then they would begin to act like animals and share their lack of morals.

Science instructors might consider telling students that there are large numbers of people of various cultures who fully accept the occurrence of evolution but do not base their morals on nature and find no logical reason to do so. Some are atheists or agnostics who have what most religious people would consider very high moral character. Others are devoutly religious, accept the science of evolution, but still base their morals on scriptures and religious traditions.

Many students quickly come to understand that people who accept evolution do not necessarily depend on nature for exemplary morals. However, those students who fully understand this may harbor another fear: a breakdown of all biblically based morals, regardless of whether people begin to act like animals.

Before students ever start to learn about evolution, many have come to believe that: (1) the foundations for the demarcation between right and wrong behavior is based on biblical scripture, and (2) the Bible is inerrant and is to be read literally (or at the least it is a God-given book that explains how humans originated in the sudden creation of Adam and Eve). Many students have not thought this through at great lengths; religious illiteracy is quite often on par with scientific illiteracy. In any case, students who hold both beliefs (1) and (2)—apparently a large number in America—fear that increasing public understanding of evolution may lead eventually to the conclusion that it did occur. This conclusion would then lead, in their perception, to the undermining of scriptural integrity in all areas, and therefore to the breakdown of all biblically based morals.

These students often are not concerned about themselves—they think they can withstand the perils of evolution education. They are concerned about the other students being taught evolutionary human origins and worry that they may not hold up so well. They do not want to see people experience an undermining of their biblical moral responsibility.

Theological instruction is most effective for students who hold such beliefs about the teaching of evolution, but it is imprudent in most public educational settings. However, the question can be answered directly in the following manner: "No, evolution doesn't necessarily lead to moral decay. Just look at all the people who find no problem with evolution and yet are upstanding citizens—many of whom are religious leaders themselves." Many go further and insist that moral values cannot be found in nature in any case—and that evolutionists make no such claim.

"Do you know about scientific creationism?"

Frequently, the student asking this question really wants to know if the instructor is familiar with how science shows that evolution is inaccurate and creationism is true. If the instructor is unfamiliar with this topic, many creationist students would be more than happy to explain it to them and supply them with literature.

The oxymoron "scientific creationism" (or creation science) was originally coined to include views of the progressives and theists, but this use proved unsuccessful.[9] It was later used by the literalists in their attempts to gain credibility outside of their own literalist theological community, particularly to gain some legal leverage against First Amendment issues—though unsuccessfully to be sure. Today the term is used almost solely by the literalists in campaigns to convince others to believe that their literalist views are not only theologically correct but also scientifically supported. Literalists Henry Morris and Gary Parker contend that the difference between scientific creationism and biblical creationism is that "The first is based solely on scientific evidence . . . the second is based on Biblical teachings."[10] Therefore, they conclude that scientific creationism should be taught in public schools.

However, courts have declared that creation science is a religious view and not science. For example, in a landmark case concerning a Louisiana statute called "Balanced Treatment for Creation-Science and Evolution-Science in Public School Instruction," the U.S. Court of Appeals, Fifth Circuit, ruled against the statute. "The statute did not protect academic freedom and attempted to discredit evolutionary theory by substituting religious instruction in the form of creationism." The case went on to the U.S. Supreme Court, where in 1987 the Court made the final ruling against the "balanced treatment" statute. "There was no legitimate state interest in protecting a particular religious view from an antagonistic scientific view."

In 1987, the Court struck down a Louisiana statute which attempted to encourage the teaching of creation science instead of the theory of evolution. The statute violated the Establishment Clause of the First Amendment because it had no clear secular purpose.

In 1981, Louisiana Legislature passed a statute called "Balanced Treatment for Creation-Science and Evolution-Science in Public School Instruction." The statute called for equal time instruction in creation science when evolution theory was taught. The statute specified the development of curriculum guides and supplied research services for creation science only. The statute's stated purpose was to protect academic freedom. A group of Louisiana parents, teachers and religious leaders challenged the act in a federal district court, claiming it violated the U.S. Constitution's Establishment Clause. They sought an injunction against Louisiana state education officials and declaratory relief. The court ruled the statute unconstitutional and the state education officials appealed to the U.S. Court of Appeals, Fifth Circuit. The Appeals court affirmed, ruling that the statute did not protect academic freedom and attempted to discredit evolutionary theory by substituting religious instruction in the form of creationism. The U.S. Supreme Court agreed to hear the case.

Applying the Establishment Clause test first described in *Lemon v. Kurtzman* . . . the Supreme Court held that the statute failed to pass constitutional standards. The Supreme Court noted that the statute called for sanctions against teachers who would not teach creation science. It did not further its stated purpose of protecting academic freedom, and had no evident secular purpose. There was no legitimate state interest in protecting a particular religious view from an antagonistic scientific view. The statute clearly had the purpose of advancing a religious doctrine and state legislatures who had enacted the statute were seeking to restructure the state's public school science curriculum to conform with the doctrine. Because the primary purpose of the statute endorsed a particular religious doctrine, it impermissibly furthered religion in violation of the Establishment Clause. The Court affirmed lower court decisions *Edwards v. Aguillard*, 482 U.S. 578, 107 S.Ct. 2573, 96 L.Ed.2d 510 (1987).[11]

Despite such significant losses and others, most literalist creationist organizations still claim that scientific creationism should be taught as if it were science—not religion. Moreover, they still contend that the scientific evidence points away from the evolution taught in most science courses and toward some form of special creation, which should be taught in the science classroom. (See Chapter 4.)

"Why would God use such a time-consuming mechanism as evolution?"

This question may be posed in many forms, such as: "Why would God use such a cruel system as evolution?" or "Why would God use such an ergonomically inefficient system to create humans?" Naturally all these types of questions are theologically—not scientifically—based. Again, as previously stated, it is usually ill advised to enter into a theological discussion with students when the ultimate educational goal has nothing to do with changing their religious beliefs but rather their science misconceptions.

However, some instructors have intentionally posed a similar question to students during class time. Instead of conducting a discussion centering on religion, they focus on hypothetical extraterrestrial beings and therefore rephrase the question to: "Why would a super-advanced extraterrestrial civilization use evolution to bring humans about on earth?" The instructors do not ask this question to dodge religiosity but rather to have a discussion about why science cannot answer questions such as these—whether specifically religious or not. Other reasons to shift away from religion include avoiding unnecessary encounters with sensitive feelings of students and helping to reduce religion-specific bias when answering the question. The instructor proceeds to have students answer another type of question: "If that advanced civilization had created humans instantaneously, essentially as they are presently, why would the extraterrestrials not have created humans with the eyes and wings of hawks so they could see better and fly, or with the stronger muscles of gorillas so they could survive better in the wild?"

Instructors then have students work in small groups to answer these questions. Usually students report back to the class with a wide variety of rationales as to why humans were designed without such benefits. The instructor then asks the class to make a *scientific* argument as to why the extraterrestrial civilization decided to create humans the way they are. It is not long before many students begin to realize that scientific arguments cannot be made about extraterrestrial design preferences for humans. By using this process teachers can show students that their question (and the question the instructor poses in our example here) simply lies outside the realm of science.

"Are you telling me that miracles don't happen?"

Generally the creationist asking this question wants to know whether the instructor believes that miracles have occurred or can occur (especially those mentioned in the Bible), and/or whether the instructor contends that science rules out the possibility of the occurrence of miracles (again with particular attention to those mentioned in the Bible). The first possible meaning of the question is usually considered very personal and of a religious nature. However, many creationist students genuinely want to know what their instructor believes about the subject. The instructor's answer to this question is of utmost importance to most Christians because it will indicate to them whether they share worldviews that supernatural occurrences have taken place. Some of the most significant supernatural occurrences to Christian creationists are the virgin birth of Jesus, the miracles Jesus performed, and Jesus' resurrection from the dead.

The importance to creationists of miracles occurring cannot be understated. Miracles are usually considered far more important and vital to creationists' lives (and afterlives) than anything they could possibly learn in school science curricula. So if the instructor chooses to answer this question, the response will probably engender one of two feelings in the student: (1) This person shares my worldview; we probably share many common values, and I can therefore relate to this person. (2) This person does not share my worldview; s/he possesses false beliefs that are of grave importance, and I therefore cannot relate to this person.[ii] The question is clearly important and relevant to many creationist students. Some of us would consider this to be an invasion of our privacy.

Some years ago one of us took a survey course on New Testament history and literature at a large secular university. The university's department of religious studies offered the course and, as is the case in most religion departments in today's large secular universities, this department's courses were not taught in an attempt to proselytize or even teach from a theistic point of view. Rather, courses were taught as mythology, important literature, or an historical examination of a past culture's stories. In fact, most religion departments at secular institutions do not exist to train religious leaders, as is commonly thought. That mission is primarily the job for seminaries and similar institutions.

The first class meeting had a full classroom with approximately 25 students in attendance. The professor began the class with the obligatory "I am so and so," "My areas of expertise and research lie in this and that," followed by a summary of what the course goals were, what would be covered, how we would be evaluated, and so forth. This first class session ended with the professor inviting students to ask any questions they might have about the course. This is where it became interesting.

Three students, two of whom knew each other prior to the class, began asking the instructor a series of pointed and poignant questions. The questions were not about the course's subject matter or procedural aspects of the class meetings, but rather were about the professor himself. The questions were all asked in a very polite manner. "Are you a Christian?" To which the professor replied "Yes." "Do you believe in Jesus' virgin birth, the miracles He performed, and His resurrection from the dead?" "No, I'm sorry I don't." "Do you believe that the Bible is the

---

[ii] Even if the instructor responds by stating something like "the question is unanswerable but interesting," the student will still have the feelings expressed in option (2) above.

inerrant, infallible word of God?" "No." "Then you're not a Christian are you?" "Yes I am a Christian and I attend church regularly." "A born-again Christian?" "No." Then one of the three students asked with the most heartfelt polite curiosity, "Why do you consider yourself a Christian?" To which he replied "I believe that the Bible is a good source of moral guidance, that the stories that it tells are about how to live a good Christian life—a moral life. The examples are what I try to follow in my life. Therefore, my holding of this philosophy makes me a Christian—a follower of a Christ-like way of life. I attend a church that shares these points of view." The inquiring students then thanked the professor, left the classroom, and never returned for any future classes. After class, some of us asked the professor if he had ever experienced such an occurrence previously. He said that such questioning from students occurred every semester but not always publicly; quite often students asked these sorts of questions one-on-one.

It would seem that students with these types of concerns would end up taking very few if any courses at a secular university because the professors teaching the courses would not pass the litmus test. Yet these creationist students take and complete the full range of classes typically offered at secular universities. The primary reason is that the vast majority of courses they take do not present material perceived by the students to be directly contradictory to some of their most cherished and fundamental biblical beliefs. Creationist students generally see no compelling need to query these professors about their Christianity. Many do not think that professors' Christian beliefs have direct bearing on the material being presented in certain courses such as physical education, for example. However, when courses present material about Jesus' life such as in a New Testament survey course, or about the origins of humans in science courses, many students perceive a possible direct conflict to their beliefs. These are some of the reasons why creationist students think it is necessary to ask science instructors questions about things that many perceive to be quite personal and irrelevant to teaching the course.

We do not recommend that a science instructor respond to such creationist students' questions by telling them that their questions are irrelevant. Such a response is really a theological argument concerning the biblical basis for such a belief, not an argument concerning the nature of the course's science content. Such a response is especially imprudent if teaching in a public K–12 school where the teachers should be remaining neutral concerning religious viewpoints. In general we recommend that teachers should not answer personal questions in the public forum. If teaching at an institution of higher education where no such constraints exist and the instructor has the desire, time, and background to take on a theological argument of this sort, he or she should realize that the discussion will ultimately not be between the science instructor and the creationist student, but will be between the instructor and theologians who have devoted their life studies to such arguments. Creationist students will seek counsel from those who will point them to publications theologians have written for the purpose of winning such arguments. Creationist theologians view the purpose of such arguments as defending Scripture and converting lost souls to Christianity.

Many instructors might scoff at the idea that their students would take the time to do any reading and research in this area as we suggest they will. Such a reaction makes sense, since many students barely do any reading for classes. However, some creationist students *will* take the time because the argument involves their most cherished and important beliefs that are much more important to them than their science courses. Even those students who will not do background reading on the subject will likely get verbal counsel from others, particularly church leaders, who have read the theological publications.

Many science instructors who decide to take the theological debate approach sometimes think that they have convinced the student of their point of view because the student no longer comes around the office to have further discussions. However, many students, during counsel with their church leaders or via publications they read, are advised to present the scriptural point of view respectfully and then stop. The idea is that if, after posing their initial argument, the instructor does not seem to be yielding to the "truth," then the students should use their time more fruitfully with others who may be more open to considering their point of view. Their consolation is that they have planted seeds that may grow in the future.

If the student perceives that a non-Christian instructor is poking fun at or being sarcastic about a belief that the student holds, this alone (whether meant intentionally by the instructor or not) may be enough for the student to give up coming to office hours for such a discussion. It may not be because the student is incapable of handling a little humor or sarcasm about what they believe, but rather because many church leaders and professional creationists believe the Bible coaches against attempting to teach biblical truths to such individuals. For example, Henry Morris in his publication *The Defender's Study Bible* gives his interpretation of Matthew 7:6, "It is counter-productive to try to teach the treasures of Biblical truth to those who reject and ridicule the Scriptures."[12] (The King James version from which he draws this interpretation is "Give not that which is holy unto the dogs, neither cast ye your pearls before swine, lest they trample them under their feet, and turn again and rend you.") When students no longer return to argue, instructors may think that students have changed their points of view due to their having presented such clear logical arguments. More likely, however, the students decided it was prudent to discontinue trying to convince their instructors, felt sorry for them, and probably now pray for their conversion regularly.

So when it comes to having a dialogue concerning the irrelevancy of the beliefs that the science instructor holds, whether about Christianity in general or about miracles in particular, it may quite often be a futile exercise. If the instructor desires not to have a debate about irrelevancy and thinks that personal religious beliefs (or a lack thereof) are no business of the person making the query, then the instructor should politely make that latter point. Because the creationist asking the "miracles" question is likely a Christian, the chances are very good that, despite an instructor's refusal to answer a question about personal religious beliefs, the student will assume that the instructor is *not* a fellow fundamentalist Christian with similar views about miracles. They will make this assumption because fundamentalist Christians rarely refuse to declare their Christianity or important Christian beliefs when asked, especially when they are asked in private. Therefore, even if an instructor has not revealed what s/he believes about miracles, the person asking the question has a high level of confidence that there is no shared worldview.

A common and usually productive response to the miracles question is that miracles certainly may occur, but they can neither be confirmed nor denied by science because science deals only with natural studies. That is, scientists study phenomena and provide natural explanations for those phenomena, not supernatural explanations. For example, a group of persons claimed to have witnessed a miracle—the liquid in a bottle turned from one color to another without any immediately apparent cause. (We are using this simple example for ease of discussion. However, it should be noted that the Bible does tell a story about Jesus performing a miracle by turning water into wine.) If the people witnessing the "possible miracle" of the liquid change wanted to know whether there was a chemical explanation for such an occurrence, they might go to a chemist to inquire. If they wanted to know whether there were supernatural explanations for such an occurrence, they might go to their clergy. If the chemist could not provide an explanation for how the change occurred, the chemist (whether or not s/he personally believed the occurrence was supernatural) would *not* provide a supernatural explanation. As a chemist, that person would continue to research the phenomenon for a natural explanation.

An analogy might be made to a police department forensic scientist who, despite a personal belief that the suspect did not commit the crime, still examines the evidence thoroughly and completely, searching for any signs that may link the suspect to the crime. Would we want that person to behave differently? It seems that most people understand that forensic scientists, chemists, and scientists in general certainly may have their own beliefs—supernatural and otherwise—but we still expect and want those professionals to examine the evidence in searching for natural explanations regardless of those beliefs.

Who decides when a miracle has occurred? Consider an event told by the immensely popular biblical apologist Josh McDowell of the Campus Crusade for Christ, who has spoken to more than 8 million students and faculty in 74 countries. (See Chapter 1, page 17.) He writes:

> During Easter week at Balboa, I first learned of the authority of the believer. About 50,000 high school and college students came down for Easter. With André Kole, the illusionist, we packed out a big ballroom. . . .
>
> As André was performing, this guy pulled up with his Dodge Dart all souped up. With a deafening sound, he popped the clutch and went

roaring down the street. Everyone inside, of course, turned around and looked outside to see the commotion. Finally, André got them settled down. . . .

When the guy went back around the block again, I knew that if he repeated his performance one more time, it would be disastrous. Turning to . . . one of our staff members, I said, "I think Satan is trying to break up this meeting. Let's step out in the doorway and exercise the authority of the believer." So we stepped outside and prayed a very simple prayer.

When the guy came back, he started to rev it up again, and as he popped the clutch—pow! The rear end of his car blew all over the street. By that time, we just thanked the Lord.[13]

Maybe a miracle did occur—the cause of the mechanical failure. But when the auto mechanic attempts to find the cause of a breakage, he will certainly look for a natural cause. If he cannot determine the cause, he does not chalk it up to a potential supernatural cause. Even those of a devoutly religious faith want to know the natural causes for why their cars won't run. Science, like auto mechanics, looks for the natural causes. The alternative would be a nonscientific attitude that, when a question cannot be answered or a phenomenon cannot be explained, simply invokes the supernatural—a miracle. (See Chapter 6, What should we do if empirical evidence and materialist philosophy are going in different directions?)

This position is usually not well received among the creationist community. Many, such as the Christian philosopher J. P. Moreland, acknowledge that while some theisms may have a god who is so capricious as to intervene frequently in the natural workings of the world, "Christian theism holds that secondary causality is God's usual mode and primary causality is infrequent, comparatively speaking." Primary and secondary causality are used here to distinguish between God's direct acts (miracles) and God's indirect acts (causation by natural laws) respectively. Moreland, as well as the majority of creationist leaders, contend that science does not nor should not defer to primary causality "willy-nilly."[14]

This brings us to the crux of the matter: Who decides what is willy-nilly? Creationists apparently want the seat of judgment to decide when something should be deemed as a primary or secondary cause, not only in

Christian theology but also in science. They apparently would want to be the ones to make the rulings on the liquid change phenomenon or the car breaking down. The creationist criteria for making a ruling of primary causality might go something like this: If

1. sufficient scientific examination (as determined by the creationists) has been conducted to determine if a secondary cause can be found for the phenomenon, and

2. enough time (as determined by the creationists) has passed for scientists to come up with new innovative ideas for what secondary causes might explain the phenomenon, and

3. enough time (as determined by the creationists) has elapsed to rule out the possibility of any future technological breakthroughs that would enable scientists to better study the phenomenon, and

4. appropriate theological corroboration exists, then the creationists would deem that the phenomenon is the result of a miracle.

Creationists rarely discuss the last criterion (appropriate theological corroboration) as an element of determining whether a phenomenon is miraculous, especially when they are discussing "scientific" creationism. The miracle must coincide with their particular theology. For example, the creationist community is widely divided on the age of the earth—from 4,000 to billions of years old. (See Chapter 2). While there is significant disagreement among creationists on how to interpret the physical data, there is even greater controversy among them on how to interpret the Bible with regard to the age of the earth. So even among the creationists themselves, there is considerable disagreement as to what would constitute appropriate theological corroboration.

The point to be made here is certainly not for us to assess which theological view is most appropriate or correct, or even to say that such an appropriate or correct determination is possible, but rather that a theological determination must be made in order to classify the phenomenon in question as either miracle or not. This theological corroboration is of great importance not only among the Christian community to insure that the non-Christian community is not assigning miracle status inappropriately, but also among factions of the creationist community, so that no faction is guilty of assigning miracle status in a fashion that is willy-nilly.

Creationists typically want to implement the other criteria of miracle status as well, not only in the creationist community but also in the scientific and science education communities. They would also hope that the scientific community would be in agreement with *their* judgments as to when adequate investigation has been conducted. However, scientists often study certain phenomena for their entire professional lives, often carrying on the research of scientists before them. Sometimes discoveries are made, and sometimes they are not. If they are not, the phenomena in question are not then deemed miraculous occurrences. They are left for continuing investigation. It appears from a creationist point of view that science cannot determine whether miracles have occurred unless theologies are also consulted, but consulting theologies is not a part of modern science—and modern science is what we teach.

8

Questions and Answers:
General Education

"Modern creationism, alas, has provoked a real battle."

—Stephen Jay Gould, Professor, Harvard University[1]

"What's wrong with presenting both sides?"

With this question, creationists are typically attempting to appeal to a sense of fairness. For decades, creationists (particularly literalists and some progressives) have told the public schools and courts that all they want is to have "both sides" neutrally taught. The "other side" they usually want presented in public school science classrooms is some version of scientific creationism. (See Chapter 2 and Chapter 7.) Creationists think that if both evolution and scientific creationism were presented equally, then students would see the merits of creationism and reject evolution as a valid explanation. Today, in addition to literalists, intelligent design creationists ask this question because they want science instruction to include supernatural causes (or nonsupernatural intelligent designer[s]).

Whether asked by advocates of scientific creationism or intelligent design, this question appeals to a democratically minded public. After all, presenting the students with both views—evolution and "X"—sounds educationally more prudent than just presenting one view. However, there are problems with such a superficial rationalization.

First, there is not just one other view; there are possibly hundreds (e.g., Hindus, Japanese Shintos, and North American tribes have creation stories). There are variations of creation stories among Christian factions, which differ in their reading of the Bible and in their church traditions. Even if those who believe that UFO-type extraterrestrials created life are not included in the mix, it still becomes apparent quickly that presenting both sides means including more than one or two additional views.[i] Presenting this information to noncreationist students or parents who think that presenting both sides sounds fair and educationally beneficial usually results in most reversing their sentiments, recognizing that there is already insufficient time to teach the existing curriculum.

Aside from issues of time for instruction, the fairness question is still faulty because it mixes apples and oranges. A federal judge has ruled creation science to be "not science" but a religious concept.[2] Therefore, presenting both sides does not mean including various scientific theories but rather including religious ideas, or at least nonscientific ideas, alongside scientific theories in science instruction. With this understanding, the fairness question becomes: What's wrong with teaching nonscience in a science class? If the general public were to hear this more precise question, the response would likely be different. Even many creation science advocates whom instructors encounter are often unaware of the variety of creation stories that exist and of the legal conclusions that have been rendered. Such persons would benefit greatly by learning such facts.

Most intelligent design proponents counter such arguments in the following way. They contend that their position is so broad that it "covers" any intelligent designer in nearly any creation story. Moreover, they assert that the legal rulings concern scientific creationism, not their intelligent design theory. For the purposes of this discussion we will concede that intelligent design theory is sufficiently broad to cover a large spectrum of creation stories. But as with scientific creationism, intelligent design theory is not in standard scientific journals because it is not science. (See Chapter 6, page 122.) So in a fairness argument, whether discussing scientific creationism or intelligent design theory, neither represents "the other side" in science. In fact, it would be unfair to students to present nonscience as science.

---

[i] By our juxtaposition of various creation stories in the same paragraph, we are not intending to imply that they are necessarily equivalent, just that they are not natural sciences.

"Why do you want to brainwash students with evolution?"

Science instructors need to explain to people who ask such questions (as ridiculous as this sounds) that brainwashing is not one of the goals of science education, but one goal is to facilitate student understanding of evolution. If some students enter a course rejecting evolution and leave accepting that evolution is a confirmed scientific explanation due to their increased understanding of evolution, that outcome would simply be a byproduct of accomplishing an educational goal of increasing an understanding of evolution. It certainly would not be due to brainwashing.

Colleagues report having success in allaying creationists' fears by replying to such a question in the following manner: "No, I'm not trying to brainwash any students, I just want them to better understand the science of evolution. I just teach; I don't force them to accept it. Whether they accept it or not is naturally up to them." Explaining that students would never be asked on a test whether they thought evolution was accurate can emphasize the point.

Nevertheless, many anxious creationists fear that students will think that evolution is justifiable merely because the instructor presents it as an accurate theory. Results of research indicate, however, that secondary students are not likely to change their beliefs in special creation due to lessons on evolution.[3] The high retention rate of student creation beliefs throughout evolution instruction is usually nothing new to those who have been teaching evolution for a number of years. They know it from experience. Reporting such findings to concerned creationists might alleviate some fears.

At the college level, instructors could point out to nervous creationist students that creationists have received advanced science degrees from major universities in various aspects of evolution. Those persons were not "brainwashed" during their schooling and they graduated even while thinking that evolution is invalid. (N.B. Only a small number of people obtain advanced degrees in areas of evolution while rejecting it, but these students nevertheless illustrate that it is possible to be exposed to evolutionary concepts over a long time and still maintain creationist beliefs.)

Concerned creationists may respond: What good is it for students to be exposed to something that they think is false even after instruction? Isn't it just a waste of valuable classroom time—time that could be better used to learn good science? In response, many science instructors find it effective to explain that regardless of whether a student deems evolution accurate, an awareness of evolutionary science can still be very helpful, and in many cases essential, to their future studies and careers. High school students going on to college will probably encounter courses that involve evolution, even if they do not major in science (e.g., a required nonmajors science course). For science majors, they would likely be required to take courses that involved evolution, even (important point to make) at fundamentalist Christian creationist colleges and universities. (Of course, evolution would be handled differently in secular universities than in those with creationist beliefs. Students attending higher education institutions with creationist tendencies will be exposed to evolution in order to learn why it is "incorrect.") So even the students who go on to creationist institutions of higher learning will most likely profit by having a better understanding of evolution. After all, it is probably helpful to first understand evolution before learning what is supposedly wrong with it.

"If science explanations are tentative, why don't we teach all explanations?"

How can a person accept something as accurate while simultaneously considering that it may be found inaccurate in the future? The creationist asking this type of question is probably attempting to weaken the instructor's case that evolution is a confirmed explanation. If the creationist can show the class that even the instructor agrees that a current scientific explanation may not be accurate in the absolute sense, then the creationist will try to show that we should open the door to teaching other possibilities as well—namely creationism.

Scientists and science instructors know that explanations in science may someday be revised or completely discarded as new evidence is discovered and presented. However, this does not mean that all current scientific explanations are equally valid. All one has to do is examine scientific journals to see competing scientific explanations and the evidence for each. Usually over time one explanation will emerge that better explains the data than the competing ones. If the explanatory power is sufficiently strong, the explanation is considered confirmed. Even in cases where the evidence is so overwhelmingly in favor of a scientific explanation that no rational thinking person could possibly consider it inaccurate, the practice of modern science still leaves the door open for new evidence that might weaken the explanation or eventually disconfirm it.

This scientific principle of tentativeness is often difficult for students to grasp—creationist or not. A review of research literature in higher education reveals that "many of our students are epistemological dualists: They view the world in rigid categories of black-white, right-wrong, and good-bad."[4] Research results also suggest that the way most instructors evaluate/test students may actively reinforce these dualistic epistemologies and concrete thinking. As such, students often believe that scientific explanations are ultimately "proven" right or wrong with a similar logic of mathematical deduction rather than a more inductive approach.[5] That is, students typically don't understand that confirmation may come from theories satisfactorily explaining data and generalizations from observations.

Many students' dualistic mindsets contribute to their difficulties in understanding how something can be confirmed while simultaneously remaining open to change. As a result, creationist students trying to point out that evolution is a wrong scientific explanation often believe that the principle of tentativeness aids them by illustrating that science should not have much confidence in the theory.

Students should be taught that biologists, geologists, paleontologists, and others show how the present evidence overwhelmingly supports evolutionary theory. The evidence is so vast that almost all of today's scientists consider evolution confirmed to the extent that they would be shocked if it were ever disconfirmed. Evolution is as highly corroborated as the notions of gravity or a round earth that rotates on its axis traveling around the sun.[6] Nevertheless, evolution is open to potential competing scientific evidence (although leading creationists will say this is not so). Creationist students should be taught that evolution is like most other highly confirmed scientific concepts in which they have confidence. Instructors should show students the similarity between their acceptance of the evidence for a nonevolutionary theory such as the germ theory of disease, the cell theory, heliocentric theory, plate tectonic theory, or the atomic theory of matter and their justification for the subsequent confirmation of it. Then instructors should ask them to justify why they have differing rules for confirming that theory and the theory of evolution. The discussion should reveal that it is probably not solely the issue of evolution rejection and tentativeness but rather a much broader issue of understanding scientific explanation in general.

"Why can't evolution be presented in an unbiased manner?"

Often the person asking this question really wants to know why evolution cannot be presented as just one of two or more equally accurate theories (refer back to the first question of this chapter). Alternatively, the person asking this question is trying to make a point about impartiality. He or she may think that the instructor, an evolutionist, cannot be impartial. This assumption negates any science instructor's claim to "fairly" present the evidence for evolution (provided of course, the instructor finds the evidence for evolution scientifically compelling).

The first thing to consider is what the person means by "biased." Meanings range from someone who is supposedly "guilty" of simply having a position, to someone who exhibits or is characterized by an extremely personal and irrational distortion of judgment. The former meaning can refer to virtually anyone who has been exposed to the topic about which they are being accused of bias. A person with a truly unbiased mind is probably rare or nonexistent. Nonetheless, even being biased by simply having a position often carries the negative connotation that one cannot be fair. The underlying assumptions of this connotation are at least twofold.

First, the definition of fair on any particular matter of judgment would be dependent on one's underpinning positions concerning the nature of that matter. One example of this is a woman's right to have an abortion. Many people with liberal religious positions or no religious positions (depending on one's definition of religious) consider it fair and legal to allow a woman to decide whether to have an abortion. Those with a conservative religious position generally do not consider it fair. Because these two camps have different underlying principles on which they base their assertions of what is fair, the result is that they both commonly contend that the other is unfair. Therefore, under these assumptions of fairness, it might be very difficult (if not impossible) to find an adult who would be regarded by both camps as fair on this issue. This scenario is often played out in real-life drama when a nomination is made for a United States Supreme Court Justice. In sum, the first underlying assumption concerning fairness is that to be fair one must hold the "correct" underlying positions that enable fairness. Thus, creationists holding this assumption will automatically negate any science instructor's claim to fairly present the evidence for evolution if the instructor holds underlying positions relevant to the issue that differ from theirs.

The second assumption is that if individuals have a leaning one way or the other concerning the issue, then they cannot be fair concerning the issue. With evolution, creationists generally make the accusation clear with statements such as: "You cannot present the evidence for evolution fairly because you are an evolutionist."

There are many ways to help students understand that people can hold differing underpinning principles or be directly opposed to something and yet be fair about the issue. Sports analogies are often useful. People understand that sports officials (e.g., referees and umpires) probably have their favorite teams and players, just like other people. Officials have spent large portions of their lives watching and learning about their game and the people who play it. Yet these officials, creationists would agree, believe that they can fairly explain and execute the rules of baseball, basketball, football, and so forth. In somewhat the same manner, science instructors can fairly explain and execute the "rules" of biology, chemistry, physics, and so forth.

Science instructors may wish to take this example further and have students role-play sports officials. The students role-play baseball umpires hired to officiate the World Series. However, they believe that the rules of the game are not correct. They think that the rules concerning unsportsmanlike behavior should be modified so that any player who curses should be thrown out of the game. After all, there are rules concerning verbal insults to the officials that result in expulsion from the game, so it would seem reasonable that there can be other word-use rules added to the book.

The questions to the students are, "Could you fairly officiate the game with the *usual* rules, not with what you would prefer the rules to be? Could you present and execute the rules of the game without misrepresentation?" If the students answer "Yes" (which is likely), then they understand a scenario in which they can hold a particular position concerning an issue and yet still be fair. Therefore, in answer to the accusation "You cannot present the evidence for evolution fairly because you are an evolutionist," the reply is "I can in the same way that you can present and execute the rules of a Dodgers vs. Cardinals baseball game even though you may not agree with all the rules—or are a Dodgers' fan."

Another approach to answering the bias question may be to ask the questioner what s/he means by *bias*. For example, the instructor may respond, "By using the word *bias,* are you asking whether I have a position about the subject?" If the answer is yes: "I do have a position about evolution being taught in the science classroom. As a good science instructor I have a position about everything that I teach in the science classroom. For example, I have a position about the validity of balancing chemical equations. I think that students should be taught that balancing chemical equations is fundamental to their chemistry education. Likewise, I find the evidence for evolution so compelling—as does the overwhelming majority of the scientific community—that I would find it incomprehensible and unconscionable not to present it to students as the scientific explanation for the diversity of life on earth. Doing less would be misrepresenting the state of modern science. Furthermore, in the same way that I found chemical equations important in my own chemistry education, I also found evolution indispensable in my own biology education."

A further response to the position issue is, "Of course I have a position. Do you expect someone who has studied a subject extensively to not have a position on it?" Quite often people making bias accusations (even including some academics) believe the myth that the scientific community, in particular, operates without positions and that scientists somehow possess pure objectivity as they go about their work. If this is what the accuser believes, then this might be an opportunity to educate the individual on some basic points in the sociology of science. Although members of the scientific community are frequently cognizant of many of their personal biases, quite often they are not, and these biases affect their work. With regard to the scientific community, Stephen Jay Gould, past President of the American Association for the Advancement Science (AAAS) writes that "without a priori preferences, we would scarcely be human."[7]

Even though scientists, like people in every occupation, are not purely objective about their work, scientists do not publish the results of their work in isolation. Students should be shown how scientists publish their methods and data, and explain their rationale for coming to their conclusions. Although there are no "science police" who examine whether scientific journal articles are overtly biased, there is a system of peer review. Scientists who share an author's particular research focus generally review the merits of a manuscript before a journal accepts it for publication. In general, journals provide reviewers with a list of criteria by which to review a manuscript and multiple reviewers examine the work and comment on its merits and faults. The reviews of manuscripts are often conducted anonymously. Authors' names are not disclosed to reviewers, nor are reviewers' names disclosed to authors. This process decreases the political effects of the review process. This overall quality control system is one manner in which scientific journals decrease the potential for inappropriate bias. It is probably safe to say that there is not a scientist living today who has been actively publishing for at least a few years who has not had a manuscript returned as unpublishable "as is."

Some students asking the bias question may really be asking if the instructor and the vast majority of the scientific community hold the position that evolution is scientifically accurate. If this is the case, one very brief and appropriate response would be, "Yes, the position I hold and that of the overwhelming majority of the scientific community is that evolution is the most accurate scientific explanation of the diversity of life on earth."

Another meaning of bias is one who exhibits or is characterized by an extremely personal and irrational distortion of judgment. This meaning of bias generally is not directed toward instructors' decisions to teach evolution but rather to the scientific community for contending that evolution is scientifically accurate for other than good reasoning. The leading creationists have produced thousands of products (books, newsletters, tracts, monographs, videotapes, and audiotapes) making such accusations. There are many responses to such charges. Because the question is not about the instructor's position concerning whether evolution should be taught, but really about whether scientists use good reasoning to come to their conclusions, it would be appropriate to direct the accuser to the scientists conducting the research (or to their publications). Generally what follows is another fallacious accusation: Because the instructor cannot or is not willing to answer the specific question posed by the accuser, the instructor is teaching something that is indefensible.

Many instructors enjoy arguing science with creationists for a variety of reasons. However, even those who enjoy such activities often have neither the expertise nor the time to research all the answers to technical questions. For example, many biology instructors may not know the specifics of radiocarbon dating or the decay of the earth's magnetic field with regard to the age of the earth. Likewise, physical science instructors may not know the specifics of mutation and natural selection. No one can have all the answers to all the questions that anyone—creationists or not—could bring up about evolution. However, this "problem" also illustrates how great and varied the evidence is for evolution. There are so many areas of science that support evolution that not even renowned evolutionary scientists have expertise in them all.

9

Questions from Instructors

"I am now more worried about the chilling effect
of creationism on teachers than I am about explicit bans."

—Donald Kennedy, President Emeritus, Stanford University[1]

The chief danger to evolution education
comes . . . from teachers just quietly ceasing to teach
evolution because it is too controversial."

—Eugenie C. Scott, Executive Director,
National Center for Science Education[2]

"Isn't it just better to de-emphasize evolution?"

This is an easy question to answer. No! Many science instructors get so much resistance from students, administrators, parents, and others about their teaching evolution that many intentionally de-emphasize evolution in their courses to avoid conflict. We can certainly sympathize with instructors who feel attacked for teaching one of the fundamentals of science. Most did not go into education to experience such a high level of discomfort.

Although many national science leaders are confident that legal attempts to ban the teaching of evolution in the United States will continue to be ineffective, they are becoming aware that grassroots efforts of professional creationists may be causing vast numbers of science instructors to either diminish their teaching of evolution or eliminate it altogether. To diminish or eliminate evolution education from the life sciences makes about as much sense as removing gravity from an introductory physics course. (See Chapter 5, "Why should students learn evolution?") Diminishing evolution education does far more harm to students than experiencing the conflict that may arise due to its teaching. Avoiding conflict only puts it off until another day, class, course, school, or election. Moreover, experiencing the conflict in learning evolution (even though it usually cannot be resolved totally in our classes) gives students an educated idea of what they are arguing when a conflict about evolution arises in the future.

"What do creationists think of me?"

Since students, parents, and other concerned parties have different personalities, backgrounds, and viewpoints, it is impossible to present accurately what an individual creationist may think of a particular instructor who teaches evolution. However, what we can report are some general trends that we have observed over the years from studying creationist products and talking to hundreds of lay creationists—the ones most often in classrooms or visiting after-class hours for a discussion.

The teaching of evolution emotionally upsets most creationists. Some are angry and confrontational; most just feel sad and are extremely quiet about it. Both groups, however, think that the instructor is misguided, uninformed, unwise, and is probably not a fellow believer. Those who are confrontational usually make it obvious that they think the instructor should be corrected, not only for the instructor's sake, but also for the sake of current and future students. They see the teaching of evolution as an attack against what they personally believe and therefore are motivated to act. They often view the instructor as a sort of academic enemy (sometimes worse) whom they must defeat. This does not mean that they show the instructor disrespect. On the contrary, creationist leaders usually emphasize that all encounters should be conducted with respect and civility. But sometimes the student is driven to confrontation in class because of an assertion the instructor makes that triggers strong emotions.

Unlike most other areas of science education that may trigger emotions, evolution seems to ignite the strongest. (See The Motivation for Attacking, page 24.)

On the other hand, the majority of students who have problems with evolution never mention these problems to their instructors; therefore, instructors never know that they have them.[i] These students often share the same emotions as the confrontational group, but think that they do not have the academic ability to reveal the instructor's "mistakes," do not like confrontation, and might worry about what the instructor thinks of them. Yet others in this group think the opposite. They think that they have the requisite academic abilities, do not mind confrontation, and do not care what the instructor thinks of them. Nevertheless, they think that it would be a waste of time to confront the instructor because the instructor and most of the students are not going to change anyway. They believe they are correct about evolution being false, but just leave the details of arguing the issue up to others—usually professional creationists. They have little doubt that if a professional creationist were present in the classroom, it would be a slam-dunk win against the evolution instructor and, of course, evolution in general.

For still other students, the whole matter is just too personal for them to discuss with anyone in an educational arena. They consider it a matter of religion or of deeply held personal beliefs, neither of which they think should be argued in secular public forums. They believe that this is just another part of life where someone (the instructor in this case) is unfortunately dead wrong, and, more than likely, they pray that the instructor will recognize his or her teaching errors and repent.

"Should I agree to debate a creationist?"

Occasionally, creationist students, particularly at the college level, will try to arrange a formal debate between a faculty member and a leading creationist. Faculty members who are not well versed in how these debates

---

[i] Some instructors think that because they use questionnaires to which students respond anonymously, the questionnaires will result in an accurate tally of student issues. But many students in smaller classes suspect that the instructor will somehow know who responded (e.g., by their handwriting) and therefore they respond "appropriately."

operate frequently see no problems in entertaining such an event. However, most science educators familiar with creationist debate tactics recommend other more productive ways for faculty to teach students about why evolution is scientifically confirmed while creationism is not.

There are many problems with the debate format. One of the most significant is that a two-hour verbal debate with a largely lay-public audience is not indicative of how the scientific community arrives at conclusions. With the audience in mind, the debate must be conducted at the level of an introductory course. At this level, debaters must address concepts that are very difficult and sometimes counterintuitive within the very limited debate time. A debate of two hours will allow each side a cumulative total of only one hour. That means that when a creationist challenges the faculty opponent to explain natural selection, transitional forms in the fossil record, beneficial mutations, carbon dating, the second law of thermodynamics, and so forth, the faculty member would literally have only one hour to teach the audience why these are confirmed ideas in science and why they support evolution. As most science instructors will attest, it is almost impossible to teach any difficult concept to an audience given only a one-hour class period.

Lack of sufficient time is one of the primary reasons why such debates are not good science education experiences but are rather presentations of extremely oversimplified reasoning, data presentation, and rhetoric designed to appeal to the audience's current sensitivities. Additionally, the inappropriate nature of the debate format works both ways. The creationists could be challenged by faculty members to explain to the audience in their 60 minutes: (1) how Noah's ark could have held two of every kind of organism on the planet, (2) why we don't find dinosaur footprints with humans, (3) why Adam's rib was necessary to make Eve, and (4) how we know that there is no life on other planets. (Most creationists believe that God did not create life on other planets.) Another interesting question might be how science should decide when something like a cold virus is intelligently designed and when it is not. (Many people once thought viruses were designed, or at least causally controlled, by evil spirits. Would creationists have attributed that to nonnaturalistic causes hundreds of years ago?) The list can go on, with thousands of such challenges for the debating creationists. The idea is to put the creationists into an equally defensive debate posture so they must attempt to make reasonable sounding explanations to the audience in a very short time. But no matter who is on the offense or defense, such debates are primarily entertainment vehicles for audiences who learn very little, if anything, about science from the experience.

Creationists benefit by debates, win or lose, because the public comes to think that creationist challenges may be credible simply because they are being formally debated in a reputable educational forum. We have never heard of an evolutionary biologist conducting a workshop on how to debate; however, creationist Duane Gish has given all-day workshops on debating creation/evolution (reporting significant attendance).[3] Our recommendation to faculty would be to put their energies into using good pedagogy to teach the compelling evidence for evolution over the duration of a course and not attempt to sway a fraction of a debate audience with sound bites. If a debate is to be held, it should be framed in such a way that evolution is not the only concept being challenged and also that the creationist alternative is being challenged equally. Naturally, this is not the arrangement under which creationists prefer to debate, but they may be willing to comply to get an audience, which seems to be their major objective.[ii]

"How can I learn more about creationists?"

Sometimes people who hear about the culture of creationists want to know more because they are shocked that so many people can hold such beliefs in our modern society. Some want to learn more because they have friends, acquaintances, or relatives that contend evolution is inaccurate and would like to challenge them. Many are science instructors who want to better understand their students' prior conceptions about evolution, creationist "science" in general, and the Christian culture from which many students who reject evolution come.

[ii] One of us has sparred with prominent creationists in front of large audiences and, consequently, the press has given the creationists coverage. For example, a newspaper reported one of these events as "Scholars Debate Creationist Theories at Harvard University" (Rechler 1999). There, we (Brian Alters and Graham Bell) formally clashed with the world's leading creationist evangelist, Ken Ham, and one of his associates. Yet while we discommend providing a forum for creationists, if such a forum will be provided to them anyway (regardless of whether opponents participate), then science and science education should be represented.

If time is available, one of the best ways to learn about creationists is via primary sources, their publications. However, we certainly do not recommend purchasing their materials since many of these organizations are partially funded by the profits they make on the sales of their books. Therefore, the purchasing of such books contributes to furthering antievolutionary activities. Many local libraries carry some creationist books, but the best way to find them is to telephone a few large local Protestant churches, Christian colleges, or seminary libraries. Noncreationists should not feel uneasy about contacting such establishments. First, many such places do not hold extreme creationist beliefs. Second, the people who work there generally are happy to help anyone gain access to their holdings because one of the primary purposes for their institution's existence is to spread the word. This is not to say that everyone will receive borrowing privileges; reading in their library may be an option. If you need to speak to a librarian to seek borrowing privileges, simply tell them the truth. For most Christian institutions, an honest rationale should be more than sufficient for them to permit access.

An excellent source of information about creationists is the National Center for Science Education (NCSE), located in Berkeley, California. They have a wide variety of types of information written by scientists and science educators about creationists. They also carry primary source materials such as creation/evolution debate transcripts, videotapes, and audiotapes. Such materials are important because creationist literature often presents only a creationist summary of a debate. NCSE also carries a book that critiques many of the creationist books, including those that are frequently suggested for classroom use. In addition, they have a bimonthly journal containing scholarly articles, news of current events, discussions and commentary, and book reviews. (See NCSE in Appendix C.)

"Why do students fight evolution education?"

Some believe that many young people just have a natural tendency toward noncompliance, and therefore this is just one of many issues in which they will challenge perceived authority—the science instructor in this case. While we do not deny that this may certainly play a role, we are not convinced that it is the primary factor. Consider that evolution is the only scientific concept with which a large number of students and parents take issue to the point of rejecting its scientific factuality. Even more telling is the fact that most creationists who attempt to discredit evolution and support creationism in public educational arenas are evangelical Protestant Christians, people who devoutly believe they should obey certain commandments in the Bible. One of the most important commandments, and one of the most familiar among evangelical Christians, is the "Great Commission." Jesus, after his crucifixion, appears in resurrected form and instructs his believers to "Go ye into all the world, and preach the gospel to every creature." (Mark 16:15) The gospel alluded to here is that Jesus was God on earth who then died for our sins and rose from the dead. People who believe this and repent of their sins, accepting Jesus as Lord of their life, will be Christians and go to heaven after death.[iii]

This commandment is given in one form or another in each of the gospels (the first four books of the New Testament—Matthew, Mark, Luke, and John), and also in the book of Acts (the book immediately following the gospels). This commandment is one of the primary reasons why evangelical Christians contend they have a supernatural mandate to spread the word to nonbelievers so that others also might become Christians. In the culture of evangelical Christianity there may be no other occurrence more pleasing than that of a nonbeliever becoming a believer—as exemplified by their use of the word *saved*. For example, they will commonly ask fellow believers if Aunt Mary or Cousin Joe is saved. A preacher or evangelist at the end of his sermon will invite the unsaved in the audience to become saved today. Saved from what? Saved from an unChristian life; saved from going to hell upon death. Naturally many (most in our experience) noncreationist Christians also share the willingness to do their part in the Great Commission, but see absolutely no connection between witnessing and people's acceptance or rejection of evolution. This

---

[iii] Many Christians will find our two-sentence explanation of what constitutes the gospel an oversimplification. However, our purpose here is not to elaborate on all the denominational nuances but simply to try to communicate the most basic concept of witnessing. Nevertheless, it is interesting to note that most of the creationists (hundreds) with whom we have discussed proselytizing and those who discuss it in print typically *do* have an understanding of the gospel that is very similar, if not identical, to that which we have explained here.

is precisely the most important point where creationist Christians disagree strongly with noncreationist Christians. The creationists contend that the concept of evolution is a major impediment—if not *the* major impediment—to a fruitful outcome of their following the Great Commission. In other words, acceptance of evolution is somewhat inversely proportional to the likelihood of considering Christianity.

Creationist assumptions and reasoning go roughly like this: Evolutionary theory suggests that (1) the days of creation in Genesis (as read by literalists) did not occur, and (2) there is no need for a creator (i.e., God). Therefore, because prospective converts have learned evolution, those people will not see the need to be beholden to the Creator or the necessity to be saved from an undesirable afterlife. This means that the teaching of evolution is somewhat analogous to the teaching of anti-Christianity or, at the very least, it erects significant barriers to the acceptance of Christianity. Christian creationist evangelist Ken Ham states:

> I've found that evolution is one of the biggest, if not *the* biggest, stumbling block to people being receptive to the gospel of Jesus Christ.[4]

Most creationists will not put it as bluntly as Ken Ham, but this is nonetheless where their primary motivations to argue against teaching evolution lie. So when students and others object to evolution being taught in the science classroom—or anywhere—it is easy to understand some possible motivations. What is surprising, though, is Christian creationists' general unawareness that the overwhelming majority of Christian seminaries have no problems accepting both evolution and Christianity. The creationist leaders certainly are aware of this fact, but often attribute it to poor biblical scholarship on behalf of the majority of seminaries. They miss the point, however. Whether it's good or bad biblical scholarship, the others still appear to be Christian—therefore evolution is not much of a stumbling block.

Another significant creationist motivation for ridding the schools of evolution instruction is their belief that without it, the world will be a much better place (as defined or perceived by creationists). Many creationist leaders seriously believe and spread the idea that

Modern ills such as homosexual behavior, abortion, promiscuity, radical feminism, and many others, likewise, look to evolution for intellectual justification.[5]

Since they find the "modern ills" they list as being greatly undesirable and believe that evolution is their underpinning, they naturally find the teaching of evolution tantamount to facilitating societal degradation. Therefore, creationists are highly motivated to bring up creationism in the classroom or to attempt to de-emphasize teaching evolution.

"Why can't creationists be comfortable with teaching that science gives one answer while religion gives another?"

Leading creationists generally think that science done "correctly" will yield conclusions that point to creationism and not evolution as the mechanism for the diversity of life. Evolution as taught in our classrooms is explicitly considered a pseudoscience by creationists. Therefore, most find no need for a demarcation between their religious beliefs and their scientific theories. Moreover, many creationists contend that science involves the discovery of truth, and the truth of science cannot contradict the truth of religion—truth cannot contradict truth. Given these presuppositions, creationists regard any attempt to relegate scientific theories and biblical revelations to different ways of knowing as an attempt to teach that science has one truth while religion has another truth. Most creationists do not believe that there are multiple ways of knowing the truth but only one—that which is revealed in the Bible. Therefore, creationists, who believe that the Bible indicates something other than evolution, believe they know *the* truth. They insist that no science instructor should try to teach others that there are various ways of knowing because there simply are not.

Another way in which science educators have tried to demarcate religion and science is by teaching that science is not about finding "the truth" but that scientific theories are about explaining the current data. All scientific theories are open to change. Quite often creationists will agree that scientific methods do not lead people to the truth but, of course, they will be quick to argue that the best scientific explanation of the diversity of life on earth at the moment is creationism. Science instructors generally follow their science/religion demarcation explanation by delineating between supernatural causes and naturalistic causes, attempting to show that advocating creationism as the best current scientific explanation is, in fact, unscientific. However, as we have discussed previously, most creationists have philosophical underpinnings that will not allow such an elimination of supernatural causation from something they "know" to have been caused supernaturally.

While the majority of Christian theologians have no significant problems with science engaging in a practice that limits itself to naturalistic causes, leading creationists often strongly disagree. In addition, students typically find it challenging to accept the "rules" of the game, especially when creationists are telling them that the rules of the game have been established, or at least modified, by unbelievers. Creationists are fond of pointing out that while we may contend that science is not about the truth, we do contend it is the truth that dinosaurs once existed on earth or that AIDS can be transmitted through unprotected sexual intercourse. They allow that science certainly does make some claims about truth. Students will also point out that many scientists do use the word *truth* in popular writings about science and will also complain when a *true* or *false* question is given on an examination (for other than pedagogical concerns). Many creationists want to put a qualification at the beginning of exams or papers concerning evolution that the following are the views of evolutionists and do not accurately reflect what they perceive to be good scientific thinking. Some creationists even would like a definition of science that includes the possibility of theories concerning supernatural creation of living and nonliving matter.

Many science instructors have found the methods of separating science and religion to be successful, but, at the same time, many have not. In those cases in which the method is unsuccessful, it is usually not that the method is unsound. Rather, it is the prior beliefs of the students and their ongoing coaching from creationists outside of the science classroom that cause misunderstanding.

"What about science teachers who think that evolution didn't/doesn't occur?"

Whether those who accept or reject evolution conduct polls of the U.S. public on various evolution topics, the results are similar. Large numbers of public school science teachers appear to question evolution's scientific factuality, and this group does not just include physical science teachers. The editor of *Science and Education* expressed his concerns when he noted a survey reporting that 30% of U.S. biology teachers reject the theory of evolution.[6]

Creationist leaders often portray the public school system as a place having so many teachers who accept evolution's scientific factuality that anti-evolution evangelism is a must. Their rhetoric suggests that there are almost no science teachers in public schools who contend that evolution is anything other than a fact. Consequently, it is surprising when creationists report their own polling data, which indicates that a considerable percentage of public school teachers do not find evolution factual. A creationist poll of those in the U.S. Registry of Science Teachers revealed that 39% disagreed with the statement "Evolution is scientific fact," with an additional 17% undecided. A total of 56% either disagreed or were undecided about evolution being a scientific fact.[7]

Moreover, many creationist science teachers apparently read the leading biology education journals. The editor of the journal of the National Association of Biology Teachers (*The American Biology Teacher*) is flooded with creationist letters when the journal publishes an anti-creation article or editorial. For example:

> In response to your issue devoted to evolution and the Scopes trial, I have sent a special donation to my favorite creation groups: Answers in Genesis and The Institute for Creation Research.

> Biology makes no sense **except** in the light of creation and a relatively short time span. I take every opportunity I can to show my students the convoluted and misleading statements and reasoning that are made in their textbooks. I show them for example that one of the best examples of evolution given in their text—the peppered moth—in fact still is a

peppered moth even now. Mutations are defects in the once perfect creation . . . Evolutionary concepts and preconceptions hamper research in biology. I want to make sure that my students are ready for the future.[8]

The situation described in this excerpt would probably not occur in other areas of science. For example, if some teachers were to teach that the earth is flat, that situation would be self-correcting—the teacher would be warned to teach that the earth is not flat. If the teacher continued to teach this inaccurate science, the teacher would most likely be reassigned out of the physical sciences or out of science altogether—not necessarily because s/he broke any law, but because of incompetence. Likewise with regard to evolution, many science educators contend that the primary concern is not what the teacher thinks is correct, but whether he or she teaches what is scientifically accurate. Whether in basic physics, chemistry, or evolution, the teacher's proficiency in having the students increase their understanding of the science is most important.

"Are creationists the only ones attacking evolution in academia?"

Certainly not. Even at the university level, concerns about the teaching of evolution are not blissfully absent. Recently at McGill University in Montreal, the Evolution Education Research Centre (EERC) was formed, with half of its professorial members from McGill and the other half from Harvard. The mission of the Centre is "To Advance the Teaching and Learning of Biological Evolution Through Research." The writer of the initial proposal naively dismissed any idea of opposition to such an educational center at the university level. Yet during the bureaucratic process of passage through numerous university committees, arguments came in from various secular universities that are surprisingly similar in intent to what the creationist leaders posit. They included the type of confused "equal time" arguments, which hold that all points of view should be taught in the classroom (e.g., that both religious and cultural views be taught in science courses). There were allegations that evolution is *only* a theory and suggestions that theologians be part of the Centre. Not surprisingly, those arguments did not come from science faculties but from other disciplines—primarily education! Even though the arguments against the teaching of evolution were from nonscience professors, they were still arguments from professors in major research universities—even institutions (ironically) with a reputation for research in evolution. While we want to make clear that such sentiments were a minority view, reports indicate that such sentiments exist at a great number of large secular universities (probably all).

The arguments against EERC stemmed primarily from religious concerns and fairness issues regarding equal time or equal presentations; however, all attacks against evolution, and science in general, do not originate from these concerns. Some academics complain that not only should nonscience views be incorporated but also that evolution and science might best be eliminated. For example, *Time* magazine reports that "Many feminist scholars . . . assert that biology is a sexist 'ideology,' not a science, and Darwin just another dead white male with an ax to grind."[9] In a book review concerning science education, Eugenie Scott, Executive Director of the National Center for Science Education, reports "there exists an anti-science movement among postmodernists that views science as negative and even 'corrosive.' "[10] Such cases against evolution, biology, or science in general are rarely furthered by students who attack evolution in science courses, and, therefore, are not discussed here due to the focus of this book—defending evolution in the science classroom. Attacks against evolution come from people in numerous camps with differing motivations. The great science of evolution must be vigorously defended in the classroom (and on all fronts) for the sake of our students' education.

10

Methods for Teaching Evolution

"Let me mention three topics that I would like
individuals to understand in their fullness . . . [one is]
the theory of evolution."

—Howard Gardner, Professor,
Harvard Graduate School of Education[1]

Regardless of whether students have taken courses in biology or evolution, they come to the science classroom with various memories, knowledge, experience, and evolution conceptions. Often, some of these conceptions are different from scientific concepts. Students develop these ideas from what they read in the popular press, find on the Web, and see on television; and from their interactions with "nature," their peers, and their parents and other authority figures. Instructors at all levels have a complex task: to tap into what students know in order to assist them in building on their scientific knowledge, to help them replace their misconceptions with scientifically useful conceptions, and to help them construct meaning from their learning experiences.

Helping students construct meaning from their learning experiences is a daunting task. Only the students can do it; instructors cannot "fill" them with knowledge. All instructors can do is to provide learning experiences that facilitate students' generating links between relevant information they already know and new information. From such a *constructivist* perspective, learning is a social process in which students make sense of experience in terms of what they already know.[2] But how does a teacher engineer lessons to help students link the new information with the old in useful, accurate, and appropriate ways? And what happens when the "old" information consists of one or more scientific misconceptions?

Many instructors know from their own classroom experience that student misconceptions (also called naive conceptions, alternate conceptions, intuitive views, or undesired understandings) are not easily changed. What many people do not realize, however, is that the process of conceptual change takes a long time, perhaps years, depending on the concept,[3] and appears to be an incremental process.[4] In learning evolutionary concepts in particular, students appear to need an extended exposure to and interaction with these concepts for growth in their understanding to occur.[5] Therefore, instructors may think of the learning experiences they provide for students as only one stepping stone toward a goal of a more complete understanding.

To facilitate a constructivist approach in the classroom, an instructor should provide situations in which students examine the adequacy of their prior conceptions, allowing them to argue about and test them. The contradictions students may face during this testing process can provide the opportunity for them to acquire more appropriate concepts. As students practice this process, they also become increasingly skilled in the procedures used in concept acquisition. More specifically, these *essential elements of instruction* are:

1. Questions should be raised or problems should be posed that require students to act on the basis of prior beliefs (concepts and conceptual systems) or prior procedures.

2. Those actions should lead to results that are ambiguous or can be challenged or contradicted. This forces students to reflect back on the prior beliefs or procedures used to generate the results.

3. Alternative beliefs or more effective procedures should be proposed by students and the teacher.

4. Alternative beliefs or the more effective procedures should now be utilized to generate new predictions or new data to allow either the change of old beliefs or the acquisition of a new belief (concept).[6]

This constructivist view is concordant with ideas on promoting conceptual change. Conceptual change theory holds that instructors must first uncover and understand students' naive conceptions before they can work effectively to help students replace these conceptions with ones that are scientifically acceptable.[7] Instructors can uncover and understand students' naive conceptions by asking students questions and analyzing their answers, or by having students solve problems requiring predictions, such as those posed in the following: "Cheetahs (large African cats) are able to run faster than 60 miles per hour when chasing prey. How would a biologist explain *how* the ability to run fast evolved in cheetahs, assuming their ancestors could only run 20 miles per hour?"[8] "The ancestor of the modern day bat could not fly, resembling a shrew or mouse. Assume that the bat evolved wings from the arm and paws of shrew-like ancestors. Explain how this could have happened using the idea of natural selection."[9] (A diagnostic test, which contains problems such as these as well as a variety of other question types, and which also includes analyses of typical student answers, is available.[10])

Students' Alternate Conceptions on Evolution

Various studies have been conducted on students' evolutionary misconceptions. Results of a study that investigated conceptions held by first-year medical students revealed that the majority of students in the sample, who had strong scientific backgrounds, left high school with a Lamarckian view of evolutionary change.[11] That is, these students thought that evolutionary change occurs as a result of need. The subjects of a subsequent study consisted of nonmajor introductory biology students, most in their junior or senior year of college. This study was designed to develop a more complete and systematic description of student conceptions about evolution. The results revealed three major ways in which college nonscience majors' conceptions about evolution differed from conceptions widely held by scientists:[12]

1. *Students thought that the environment (rather than genetic mutation, sexual recombination of genes, and natural selection) causes traits to change over time.* They failed to distinguish between the separate processes responsible for (a) the appearance of traits in a population (which originate by

random changes in genetic material, e.g., mutation and sexual recombination) and (b) the survival of such traits in the population over time (natural selection). Students had difficulty understanding the key idea that the environment affects the survival of traits *after* their appearance in the population. Instead, they thought that change was the result of environmental forces alone, which acted on organisms to produce change, and they held various conceptions about how the environment exerts its influence.

- Some held teleological/Lamarckian conceptions; that is, students attributed evolutionary change to need. For example, students thought that if cheetahs needed to run fast for food, nature would allow them to develop faster running skills.

- Some held other Lamarckian views, attributing evolutionary change to use and disuse. For example, students thought that if cave salamanders did not use their eyes for many generations, their eyes would become nonfunctional.

- Students attributed evolutionary change to organisms' ability to change in response to environmental "demands." For example, students thought that polar bears adapted to their environment by slowly changing their coats to white.

2. *Students did not view genetic variation as important to evolution, even though such variation is essential to evolution taking place.* Students viewed evolution as a process that acts on the species as a whole, not recognizing that it is variation among individuals within a population that constitutes evolution's raw material. They did not understand that the process of natural selection is dependent on differences in genetic traits and in reproductive success among individual members of a population.

3. *Students viewed evolutionary change as gradual and progressive changes in traits, rather than as a changing proportion of individuals with discrete traits.* Students did not recognize that traits become gradually established in a population as the proportion of individuals possessing those traits grows with each succeeding generation. They believed that all individuals change slowly over time. For example, students thought that if salamanders living in caves did not need sight, that they would pass down genes conferring less and less ability to see until the salamander populations were blind. In

an example from another study, students harboring misconceptions about pesticide resistance in insects thought that individual insects became *more immune* to pesticides over the lifetime of each, rather than *more insects* becoming immune to pesticides over many generations.[13] Students harboring such misconceptions do not understand that a *population* may change over generations because the proportions between individuals carrying one or another trait is altered.

In a Brazilian study,[14] students viewed evolution as a progressive change in traits, which they interpreted as progress, improvement, and growth. They viewed evolution as a ladder, with viruses at the bottom of this ladder and humans at the top. Since the ultimate "goal" of evolution was achieved—the evolution of humans—they perceived that evolution was no longer taking place.

Students' Confusion of Scientific Words with Everyday Words

Research results also indicate that *students' reinforce their naive conceptions by confusing the scientific terms* adapt/adaptation *and* fitness *with the everyday terms.*[15] This is certainly not surprising, since many scientific words are used in day-to-day language but carry meanings that differ from their scientific meanings. The word *animal,* for example, has a much less precise meaning to children as they grow up and interact with their environment than it has in a scientific context. The meaning of this word is embedded in children's long-term memories and is used as a base to construct meaning when they take science courses.[16] Unless the meaning they hold for this word is examined and compared to the scientific meaning, and unless they replace their intuitive conception with a scientifically correct conception of an animal, they will construct new knowledge with this faulty concept. The learning that results may not coincide with what instructors expect to happen!

In an evolutionary context, the meaning of *adapt* and *adaptation* refers to changes in populations of organisms that occur because of random mutation or sexual recombination of genes. Those individuals within populations that possess physical, behavioral, or other attributes well-suited to their environment are more likely to survive than those that possess physical, behavioral, or other attributes less suited to their environment. The survivors have the opportunity to pass on their favorable characteristics to their offspring. These characteristics (adaptations) are naturally occurring heritable traits found within populations. In its everyday sense, the term *adapt* refers to individuals changing their physical, behavioral, or other attributes in response to an environmental condition. The change is an adaptation. For example, a night owl student adapts to getting up early for an 8:00 A.M. class. This trait is not heritable; it cannot be passed on to another generation. This confusion of terms reinforces the misconception of students that the environment somehow acts upon individual organisms to force them to change certain characteristics or perish.

In its everyday sense, the term *fitness* refers to an organism's health, strength, or intelligence. For example, individuals are considered "fit" if they eat a balanced diet, are an appropriate weight for their height, and exercise. The definition of *fitness* in the context of evolution, however, generally refers to reproductive fitness—the ability of an organism to survive to reproductive age in a particular environment and produce viable offspring. Genes that confer a reproductive advantage increase an organism's fitness.

Addressing Alternate Conceptions in the Classroom

So how can definition confusion and misconceptions be addressed in the classroom? Determining the roots of these problems can be helpful so that specific points of faulty reasoning can be discussed with students. One researcher suggests that faulty student patterns of reasoning regarding evolution appear "to grow from an initial incorrect observation that individuals can change their characteristics during their lifetime *and that this acquired change is passed genetically on.*"[17] Further, she suggests that this "intuitive" scientific reasoning exhibited by students "mirrors a kind of 'recapitulation theory' of the history of scientific thought."[18] Others concur, noting that "novice learners frequently express conceptions consistent with earlier versions of a science concept accepted by scientific communities in the past, and frequently for similar reasons."[19]

These ideas suggest that students must receive practice distinguishing between the concepts of inheritable characteristics and acquired (nonheritable) traits. Also, students need practice in probabilistic reasoning; genetics problems used to teach Mendelian genetics that show inexact proportions of offspring help students understand the probabilistic nature of the results.[20] Additionally, instructors may find it useful to help students identify their conceptions and compare them to long-discarded scientific hypotheses that are no longer useful in solving evolutionary problems.

Frustratingly for instructors, students often leave science courses holding the same misconceptions about evolution as when they entered them, even though they listened to many lectures on evolution and genetics and received acceptable grades on exams. Why does this occur? One reason may be that presenting information about evolution in lectures promotes rote learning of these concepts and does not help students use the concepts and understand their utility.[21] Additionally, lectures alone may be inadequate to create sufficient conflict in students' minds to alter their existing understanding. Students therefore develop an "isolated" knowledge structure from this rote learning, connecting it only weakly with the knowledge structure they already hold.[22] When the class is over and they are back in the "real world," they return to their intuitive views (their old ways of thinking),[23] and the knowledge structure developed for use in science class is forgotten over time.[24] To combat this problem and to work toward conceptual change, researchers recommend that instructors engage students in problem solving and applying concepts to real-world situations.[25] Instructors must guide students to think about and discuss the meaning of ideas and their applications in order to facilitate conceptual change and student learning, rather than simply telling students "facts" and expecting them to learn the "right" answers.[26]

Summarizing what we have discussed thus far in this chapter, to facilitate conceptual change, to promote learning in a constructivist manner, and to help students comprehend evolution concepts, instructors must first uncover and understand students' naive conceptions about evolution. This can be accomplished by analyzing students' answers to evolution problems or questions. The next step is to provide learning situations that help students examine their alternate conceptions and, ultimately, become dissatisfied with them. (See *essential elements of instruction* steps 1 and 2, p. 180.) These steps might involve focusing students' attention on observations, data, or problems that cause them to realize that their existing conceptions do not explain the observations, or are not useful in solving the problems. During this process, instructors must help students think through their dissatisfaction and realize why their naive conceptions are not "working." Instructor–student discussion is key. Such interaction should be engineered to challenge students to think and talk about their ideas, while guiding students to see the relationships and contrasts between their own ideas and the ideas widely accepted by scientists. (See *essential elements of instruction* step 3.) Finally, help students examine these new ideas by focusing their attention once again on observations, data, or problems to see if these ideas are useful in explanatory or problem-solving contexts in which their prior conceptions were not helpful. (See *essential elements of instruction* step 4.) Various teaching methods appear beneficial in accomplishing these goals.

Analogical Reasoning.   Analogies are valuable teaching tools, and most instructors use them all the time. A teacher simply explains a *target* concept by using an analogous concept (*base* or *anchor*) that is better understood by the student and pointing out how the target and the base are like one another (and possibly how they differ). For example, ask students a question that is likely to result in their stating a misconception. One student's answer to the cheetah question cited previously revealed the following misconception: "[Cheetahs evolved the ability to run fast] because they (cheetahs) needed to run fast for food, so nature allowed them to develop faster running skills."[27] Here, the student explicitly states a teleological misconception, suggesting that organisms develop new traits because they need them to survive. An instructor could use an analogy to help students realize the misconception inherent in this answer. For example, an instructor might suggest that the track team of their school or university could be thought of as analogous to a population of cheetahs. If the students need to run fast to win the regional championship, would nature allow them to develop faster running skills because of this need? Or would they run faster as a result of proper training and coaching? Would the future sons and daughters of these track team members be able to run faster as a result of their parents' training? Using simple analogies such as this help students understand that the "need" to run fast does not confer the ability to run fast, and that acquired traits are not heritable traits.

Analogies are only one way in which to challenge students' misconceptions and present students with new, scientifically acceptable conceptions. There are a variety of ways to accomplish this task, but teacher talk must be responsive to what students are saying. In order for students to adopt new conceptions, they must understand these scientifically valid conceptions, think that they make sense, and find them useful in solving problems or providing explanations in a variety of situations.

Problem Solving and History of Evolutionary Thought.   This approach incorporates many of the research-based ideas mentioned thus far.[28] First, this method assesses students' prior knowledge of the subject; it then uses a combination of problem-solving activities (using evolution problems such as those on p. 181) and discussions of historical vignettes to attempt to change student misconceptions. The use of both a historically rich curriculum and problem solving as a method of improving students' reasoning skills regarding understanding the processes of evolution is supported by various researchers[29] and *Benchmarks for Science Literacy* produced by the American Association for the Advancement of Science.[30]

Results of a study involving nonmajor introductory college biology students revealed that those who were less skilled in reasoning were more likely to hold nonscientific beliefs than were students more skilled in reasoning. They noted that nonscientific beliefs held by less-skilled reasoners are not easily altered by instruction, "thus instruction should focus on ways of improving student reasoning skills as well as teaching scientific conceptions"[31]

The problem-solving/historical vignette approach involves:

1. asking students to solve evolution problems,

2. having students circle all teleological phrases or words in their answers,

3. asking students to solve the problems again without using such phrases or words,

4. introducing historical evolutionary theories such as those of Lamarck and Darwin,

5. modeling evolutionary problem solving using these theories, and

6. asking students to work in pairs solving problems using four evolutionary models, and to critique their answers.

(A more complete, yet simple explanation of this approach is available.[32]) The results of research indicate that this approach "produced improvement in students' understanding of evolution as contrasted with traditional instruction; but the evaluation data also indicated that much more improvement can be made." Additionally, "students will generally increase their use of Darwinian ideas, but it is a much more difficult task to reduce their use of non-Darwinian ideas."[33]

Another example of the use of problem solving and a historically rich curriculum is a course developed for biology student teachers.[34] The course first assesses student teachers' preconceptions about evolution by asking them evolution questions. After learning the history of evolutionary thought, students analyze their own responses to the evolution questions, reflecting on their responses and discussing similarities between their preconceptions and ideas in the historical development of evolutionary thought. The student teachers then read studies on children's evolutionary misconceptions and conduct research projects on misconceptions regarding evolution among school children or among other student teachers.

An evolution teaching module has also been developed that includes the use of problem solving, student–teacher discussion, and laboratory activities.[35] These modules were developed to foster conceptual change after the diagnosis of student misconceptions. Results of research on the implementation of these materials show low to moderate success in improving students' understanding of the evolutionary process.[36]

Inquiry. Learning by inquiry involves student investigation and usually includes science process skills such as formulating questions, developing hypotheses, designing experiments, analyzing data, and drawing conclusions. The *National Science Education Standards* defines inquiry learning as "a set of interrelated processes by which scientists and students pose questions about the natural world and investigate phenomena; in doing so, students acquire knowledge and develop a rich understanding of concepts, principles, models, and theories."[37]

One inquiry approach to teaching evolution has been developed by the Biological Sciences Curriculum Study (BSCS), a nonprofit organization founded in 1958 that develops and field-tests science curricula.[38] These twenty-one activities and accompanying videodisc use a guided inquiry approach and focus on discussion as an important instructional strategy, providing opportunities for students to interact with the teacher and with their peers. Studies conducted on the effectiveness of these inquiry activities showed that students significantly increased their use of scientific conceptions of evolutionary processes.[39]

The *National Science Education Standards* includes an example of an inquiry lesson that uses brachiopod fossils to help students distinguish between variation within a population, and variation versus evolutionary change among populations.[40] Students' ultimate task is to determine whether two populations of organisms are the same species or different species by collecting and analyzing data from fossil populations and reviewing the scientific literature.

Concept Mapping. A concept map is a visual representation of the hierarchical organization of concepts, showing their relationship to one another. Concept mapping forces students to establish and create relationships among concepts, which can lead to improved student understanding and can help instructors monitor the development of understanding.

To create a concept map, a person must identify the key concepts of a topic, arrange these concepts from general to specific, and relate them to each other in a meaningful way.[41] A scientific concept can be defined as "regularities in objects or events designated by some label, usually a term. Whether a process (e.g., precipitation), a procedure (e.g., titration), or a product (e.g., carbohydrate), concepts are what we think with in science."[42] In a concept map, scientific concepts are linked to form propositions about the concepts on the map. The following examples of propositions show a scientific concept first and last, connected by linking words: vertebrate embryos have notochords, environmental factors can influence survival, and mutations are a source of variation.

In a concept map, the concepts are written in lowercase letters and generally each is placed within a circle or an oval. The mapper places the superordinate concept at the top of the page, with subordinate concepts flowing down the page from least specific near the top, to most specific near the bottom. Lines and linking words connect the concepts. Generally, the links are vertical, with each branch of the map able to be read from top to bottom. Examples may be included anywhere on the map, and are designated by being enclosed in a dotted circle or oval. Dotted lines are also used for horizontal cross-links of one concept with another. Using an arrowhead shows that the proposition is not bi-directional. Figure 10-A is an example concept map on the design of a pencil, which shows most of the graphic conventions employed in concept mapping.[43]

One example of the use of concept maps in a course on evolution is discussed in a study in which students were able to develop concept maps of class lectures, one per week for ten weeks, for extra credit.[44] Students were first given training in developing concept maps. For each lecture students could map for credit, the instructor wrote five seed concepts on the board that students had to use in their maps. The students were to add five additional concepts and/or examples to their maps. The concept maps were collected at the next lecture and corrected by the instructor. Maps were kept small, and the instructor was able to correct each one in five minutes. Students spent an average of 48 minutes constructing the maps and most reported an increase in their total study time as a result of concept mapping. Moreover, constructing concept maps forced students to study throughout the course rather than just before exams. There was also a positive correlation between the quality of students' concept maps and their academic success in the course. Additionally, the instructor was able to identify critical junctures in the students' learning of evolution by analyzing the concept maps submitted after each lecture. By noting the conceptual difficulties that arise during the course, instructors can adjust their teaching approaches or the content of the course to better meet student needs.[45]

The Learning Cycle.   The learning cycle is a model of teaching in which students first explore concepts of the lesson via hands-on activities. These activities are engineered to help students develop an understanding of the concepts on which the lesson focuses before these concepts are overtly stated. After this initial exploration phase, the concepts are usually introduced using didactic methods such as lectures, demonstrations, films, or textbook readings. Students are then provided an opportunity to apply

FIGURE 10-A

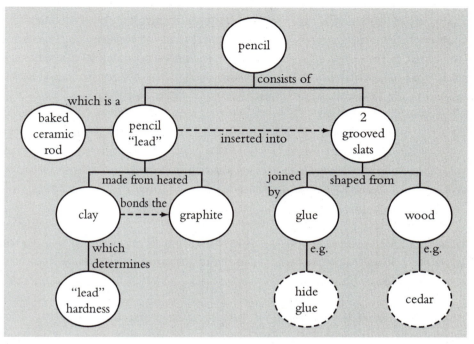

SOURCE: Wandersee, J.H. (1990). Concept mapping and the cartography of cognition. *Journal of Research in Science Teaching.* 27: 923-936 p.933.

what they have learned, with the goal of using their new understanding in other situations. This application phase may flow into the exploration of a new concept, hence the name learning *cycle*. To summarize, the three phases of the learning cycle are *exploration*, concept *introduction*, and concept *application*.

Three types of learning cycles have been proposed: descriptive, empirical-inductive, and hypothetical-deductive. All three follow the basic plan outlined in the previous paragraph. However, descriptive learning cycles focus on observation and the description of patterns occurring in nature. Empirical-inductive learning cycles require students to explain phenomena, while hypothetical-deductive learning cycles require that they test alternative explanations. An extensive description of these learning cycle models is available.[46]

An extension of the learning cycle contains five phases and is referred to as the 5E Instructional Model: Engagement, Exploration, Explanation, Elaboration, and Evalution.[47] During the engagement phase, the teacher focuses the student on a problem or situation by using strategies such as asking questions or presenting a discrepant event. The three learning cycle phases described previously follow. In the evaluation phase, students receive feedback on their understanding or are assessed in traditional ways such as tests.

Examples of the use of the learning cycle in teaching evolution are available in the literature; we cite only three here.[48] One example shows how to use an excerpt from *The Flamingo's Smile* by Stephen Jay Gould in a learning cycle activity that focuses on the idea that form follows function and the concept of adaptation. Another example shows how peer discussion can be used within a learning cycle format to explore questions about the teaching of evolution versus creation in public school science classes. A third example shows how to help students develop a sense of geological time by developing a simulated timeline in the school hallway using a roll of toilet paper.

Hands-on Activities, Laboratory Investigations, and Field Trips. The theory of evolution and the concepts inherent in this theory are abstract and are often difficult for students to grasp without concrete representations. Instructors can use, for example, time lines of a given distance to represent a unit of time or displays that allow students to see a million objects.[49] They can take students on field trips to "see, touch, and experience the remnants of past environments."[50] Students can measure and study the variability in the student population of their classroom, a few classrooms, or their school, using such traits as height, reaction time, or foot size. They can conduct a variety of laboratory exercises that include the simulation of genetic drift or natural selection, and the preparation and analysis of karyotypes of local animals and plants.[51] Many more ideas can be gleaned from the references given in this paragraph and in the list of activities described at the end of this chapter.

Multiple Intelligences

In 1983, Howard Gardner outlined a "theory of human intellectual competences" in *Frames of Mind: The Theory of Multiple Intelligences*.[52] This synthesis of ideas has evolved over nearly two decades and is an important educational idea of the twenty-first century. The theory of multiple intelligences holds that intelligence is more than a single property of the human mind. The most recent definition of "an intelligence" is "a biopsychological potential to process information that can be activated in a cultural setting to solve problems or create products that are of value in a culture."[53] The eight intelligences are linguistic, logical-mathematical, musical, bodily-kinesthetic, spatial, interpersonal, intrapersonal, and naturalist.

In the context of the multiple intelligences theory, Gardner suggests that instructors engage students using "entry points" that roughly align with the intelligences.[54] Some entry points are:

- *Narrational*—engages students who enjoy learning through stories. For example, develop lessons that use the story of Darwin's voyages or the life course of a particular species.

- *Quantitative/Numerical*—engages students who are intrigued by numbers. For example, develop lessons that look at numbers of individuals of a species in varied ecological niches and at how those numbers change over time.

- *Logical*—engages students who think deductively. For example, a syllogism that describes the concept of survival of the fittest is:

  If there are more individuals/species in a territory than can be supported, and

  If there are variations among individuals/species,

  Then those variants that survive best in a particular ecology will be able to reproduce and flourish there.[55]

- *Hands-on*—engages students who enjoy manipulating materials and carrying out experiments. For example, have students breed generations of Drosophila to observe the incidence and effects of genetic mutations.

- *Social*—engages students who learn best in an interactive group setting. For example, give a group of students a problem to solve such as determining what happens to various species in a given environment following a dramatic change in climate. The use of role-play engages these students as well.

Activities

The conclusion of this chapter comprises a listing of teaching activities. We do not evaluate these activities, but include them as a database of ideas from which to build lessons using the research-based suggestions described in this chapter.

Despite the levels for which the lessons that follow were originally designed, most can be adapted to meet the instructional needs of a wide variety of students. The original journal articles are cited in the notes and references. These articles provide further exploration of each activity. .

Topic

Evolutionary (Phylogenetic) Trees

Concept: Evolutionary trees depict patterns of relationships among major groups of organisms; they illustrate lineages based on evolutionary relationships.

• Brewer describes simple tree construction problems that allow students to become familiar with the processes used by scientists to explain evolutionary history and develop a tree-based (branching) conception of phylogenetic biology, rather than a ladder or straight line conception.[56]

• Gendron describes a laboratory activity that teaches evolutionary concepts by constructing phylogenetic trees. The activity uses "caminalcules," which are imaginary organisms with a fossil record. One of the main goals of the activity is to illustrate the connection between the classification of living species and their evolutionary relationships.[57]

Topic

Transitional Forms

Concept: Evolutionary transitions that leave signs in the fossil record.

• Hazard describes a classroom demonstration using the overhead projector to teach students about transitional forms in the fossil record.[58]

Topic

Microevolution

Concept: A change in the genetic makeup of a population; a change in the relative frequencies of alleles (genotypic ratios) within a gene pool over successive generations.

• Welch describes a class activity that simulates the process of microevolution and helps students visualize the concept of change in the gene pool of a population over time. This activity also helps students understand that populations, not individuals, evolve. Students use a "population" of beans of various colors as mock prey. They "capture" prey in grassy areas over a specified time, then calculate the number of "survivors" of each color by subtracting the number of "captured" beans of each color from the total. After students add beans to the environment to simulate reproduction by those left, they capture prey and count the remaining beans again, over successive "generations." Students plot results using a spreadsheet, which shows the shift in colors within the gene pool.[59]

• Brewer and Zabinski describe simulation activities using beans as in the Welch activity above. However, the Brewer and Zabinski activities are designed to be used in large lecture hall classes as well as in smaller classes. They use a cooperative learning format, which is well described in the article, and include simulations of gene flow, founder effect, and genetic drift.[60]

Topic

Speciation and Conservation of Genes

Concepts: New species arise when organisms that were previously able to breed with each other become reproductively isolated. Members of different species share common genes.

• In two related articles, Offner presents background data and examples of chromosome comparisons to show how to teach speciation and conservation of genes using chromosomes.[61]

Topic

Using Skulls to Teach Evolution

• Gipps describes an activity in which students are presented with skull casts of humans and their ancestors. (The particular skulls Gipps uses are listed in the article.) The students are asked to determine which skulls are human. After students suggest answers and reasons for their answers, the instructor lists the criteria for "humanness." The students then use these criteria to arrange the skulls in evolutionary "order," from *Australopithecus* to modern humans, noting the increasing human characteristics and the decreasing ape-like characteristics.[62]

• Riss describes activities, drawn from the *Stones and Bones* program, to teach students techniques used in physical anthropology to measure the condylar index of various species of animal skulls. *Stones and Bones* is a hands-on physical anthropology program for high school students developed by the Los Angeles Unified School District, the L.S.B. Leakey Foundation, and the Los Angeles County Museum of Natural History. Riss has adapted these activities for use with photocopies of skulls, which are reprinted in the article. [63]

Topic

Fossils

Concept: Fossils are geological remains, impressions, or traces of organisms that existed in the past.

• Dolph and Dolph describe an activity in which students analyze, identify, and prepare high-quality specimens of bony fossil fish. The fossils, obtained from a supplier, are from the Green River Formation (Wyoming/Utah). They are embedded in a limestone matrix, which students remove slowly and carefully using a dissecting needle and a gum eraser. After spraying the fossil with acrylic plastic to protect it, students study the fish's morphology, comparing it to illustrations of 3 possible species.[64]

• Matthews describes an activity that first evaluates students' misconceptions about fossils by having each student pick an object from a box of fossils and nonfossils and explain why they think the object they picked is (or is not) a fossil. After engaging in an instructor-led discussion about which objects are fossils and why, the class develops a definition of the term *fossil*. Groups of students are then each given a bag of 15 fossils. (Each bag contains the same fossils.) Each group picks a fossil to discuss with the class. After these initial discussions, the students categorize and study the 15 fossils using reference materials. In a culminating activity, students go on a fossil collecting trip, identify the fossils they find, and clean and mount them.[65]

• Platt describes how to develop a collection of fossil materials for use in the classroom. He discusses the types of fossils recommended for purchase and provides an annotated dealer list.[66]

• Shaw, Crocker, and Reed describe activities that provide students with the sense of patience and skill required to extract and reconstruct fossils. After the teacher presents a lecture on the work of paleontologists, students remove chocolate chips from cookies with and without a toothpick "tool," and reconstruct the original shape of large teacher-made cookies from the pieces.[67]

Topic

Adaptations

Concept: Naturally occurring heritable traits found within populations that increase reproductive fitness.

• Guerrierie describes an activity on beak adaptations in which stations are set up around the classroom with each station having a different type of bird food. Students use kitchen and office tools (such as tongs, a slotted spoon, and a staple remover) to "feed" at each station for a prescribed amount of time. They gather and compare data, and then examine pictures of bird beaks to determine which type of food the "beaks" are adapted for gathering.[68]

Topic

Natural Selection

Concept: Organisms having adaptive traits survive in greater numbers than those without such traits.

• In this activity by McCarty and Marek, students plate water samples from untreated sources (such as a farm pond or a drainage area) onto three types of media: media that selects for Gram negative organisms (allows only Gram negative organisms to grow), this same media including an antibiotic that kills Gram negative organisms, and nonselective media. After incubating the plates, students sketch the various colonies growing on the plates and perform Gram stains. Pooling class data, students see that only Gram negative bacteria grew on the selective media, that no bacteria (or only a few colonies) grew on the antibiotic plate, and that both Gram positive and Gram negative bacteria grew on the nonselective media. Teacher questions guide a class discussion regarding selection.[69]

• Lauer presents a variety of activities using jelly beans and a student's sense of taste as learning tools. He describes activities that illustrate directional, disruptive, and stabilizing selection; competitive exclusion; sexual selection; optimum foraging theory; and Hardy Weinberg theory.[70]

• This activity by Peczkis is meant to simulate natural selection working on random mutations over many generations. Its purpose is to help students understand that natural selection is not purposeful (to dispel the Lamarckian view). Students are given directions on slips of paper. Each direction tells them to darken certain squares on graph paper. Each student's "graph designs" look different from every other due to the random order of their directions. The teacher identifies certain "adaptations" (such as circular shape) that are advantageous. Designs with this shape reproduce (are photocopied) and given to students. The process is repeated using the photocopies as a starting point for the next generations.[71]

• Knapp and Thompson describe how to simulate natural selection using playing cards. Students view changes in a "gene pool" of cards over successive "generations" of saving winning hands and discarding losing hands. The authors also describe five game variations to simulate what happens to a highly beneficial gene in the population, a dominant lethal mutation, a recessive lethal mutation, genetic drift, and major environmental alterations.[72]

• Chandler asks students to assume the role of scientists living 2 million years in the future. Students choose an animal alive today, describe some of its present-day adaptations, and then describe environmental pressures involved in the selection process for the future form. The students create a "missing link" organism (poster or model) that ties the animal of the future with its ancestor of today. All is presented in a classroom evolution symposium.[73]

• Goff has developed an activity in which students work in groups to design paper airplanes whose construction simulates mutation. The initial airplanes that students construct can have only 5 folds. Similar planes are grouped as members of the same species. Students then put 3 more folds in their planes and see whether these "mutations" have increased, decreased, or had no effect on the plane's ability to fly.[74]

Topic

Museum Exhibits

• Students build models and develop their own museum exhibits about the history of life in a project described by Oden. This student research project culminates in a school museum night.[75]

• The Museum of Paleontology at the University of California at Berkeley has created an online virtual museum. Scotchmoor describes this museum, which was still online in March, 2001 at http://www.ucmp .Berkeley.edu.[76]

Topic

Law-related Teaching
of the Evolution/Creation Controversy

• Morishita describes an activity that helps students understand how United States constitutional law impacts the teaching of creation and evolution in American classrooms. First, students are assigned a pre-essay to determine their prior knowledge regarding both topics. Students read about the Scopes trial from the text *Great Trials in American History*, discuss this trial, and then prepare court cases of their own as if the trial were taking place today. Students read reviews of more recent evolution/creation court cases as part of their preparation.[77]

• Duveen and Solomon describe a role-play activity called *The Great Evolution Trial*, and discuss the rationales inherent in its development. The role-play is about the publication of *The Origin of Species* and a subsequent trial to determine if the work is blasphemous. The article describes one goal of role-play as student acquisition of empathy with characters in other times or other places, and student understanding of contemporary social issues from different perspectives. The role-play activity delves into both religious and scientific opposition to Darwin's work in the 1800s.[78]

Topic

History of Science and Evolutionary Thought

• Gauld suggests using an incident that occurred in Oxford, England, in 1860 between Samuel Wilberforce (the bishop of Oxford) and Thomas Huxley regarding *The Origin of Species* to enable students to understand the cultural context of the nineteenth century in which debates about evolution were taking place. The incident happened at the annual meeting of the British Association and reflects a wider debate in the scientific community at the time.[79]

Topic

Evolution versus Special Creation

• Lawson presents a lesson, which has been designed using a learning cycle format, that allows students to confront the evolution versus special creation issue by gathering evidence from the fossil record, analyzing it, and then deciding for themselves which explanation represents the best answer to the question of what caused present-day species diversity.[80]

• Seaford describes a method he used, which includes readings and five writing assignments, to have students explore the beliefs and goals of "scientific creationism" and contrast them with the concepts inherent in evolutionary theory.[81]

A

Organizations That
Support Evolution Education

Here are some North American scientific organizations that have attested to the fact that evolution is scientific and support its teaching in science courses. None of these organizations support creationism being taught in public school science courses.

Academy of Science of the Royal Society of Canada
American Anthropological Association
American Association for the Advancement of Science
American Association of Physical Anthropologists
American Astronomical Society
American Behavior Society
American Chemical Society
American Geological Institute
American Geophysical Union
American Institute of Biological Sciences
American Physical Society
American Psychological Association
American Society of Biological Chemists
American Society of Naturalists
American Society of Parasitologists
Ecological Society of America
Genetics Society of America

Geological Society of America
National Academy of Sciences
Paleontological Society
Society for Amateur Scientists
Society for Molecular Biology and Evolution
Society for the Study of Evolution
Society of Systematic Biologists
Society of Vertebrate Paleontology

B

Eight Significant Court Decisions Regarding Evolution/Creation Issues

1. In 1968, in *Epperson v. Arkansas*, the United States Supreme Court invalidated an Arkansas statute that prohibited the teaching of evolution. The Court held the statute unconstitutional on the grounds that the First Amendment to the U.S. Constitution does not permit a state to require that teaching and learning must be tailored to the principles or prohibitions of any particular religious sect or doctrine. (*Epperson v. Arkansas* (1968) 393 U.S. 97, 37 U.S. Law Week 4017, 89 S. Ct. 266, 21 L. Ed 228)

2. In 1981, in *Segraves v. State of California*, the court found that the California State Board of Education's *Science Framework,* as written and as qualified by its antidogmatism policy, gave sufficient accommodation to the views of Segraves, contrary to his contention that class discussion of evolution prohibited his and his children's free exercise of religion. The anti-dogmatism policy provided that class discussions of origins should emphasize that scientific explanations focus on "how," not "ultimate cause," and that any speculative statements concerning origins, both in texts and in classes, should be presented conditionally, not dogmatically. The court's ruling also directed the Board of Education to disseminate the policy, which in 1989 was expanded to cover all areas of science, not just those concerning issues of origins. (*Segraves v. California* (1981) Sacramento Superior Court #278978)

3. In 1982, in *McLean v. Arkansas Board of Education,* a federal court held that a "balanced treatment" statute violated the Establishment Clause of the U.S. Constitution. The Arkansas statute required public schools to give balanced treatment to "creation-science" and "evolution-science." In a decision that gave a detailed definition of the term "science," the court declared that "creation science" is not in fact a science. The court also found that the statute did not have a secular purpose, noting that the statute used language peculiar to creationist literature in emphasizing origins of life as an aspect of the theory of evolution. While the subject of life's origins is within the province of biology, the scientific community does not consider the subject as part of evolutionary theory, which assumes the existence of life and is directed to an explanation of how life evolved after it originated. The theory of evolution does not presuppose either the absence or the presence of a creator. (*McLean v. Arkansas Board of Education* (1982) 529 F. Supp. 1255, 50 U.S. Law Week 2412)

4. In 1987, in *Edwards v. Aguillard*, the U.S. Supreme Court held unconstitutional Louisiana's "Creationism Act." This statute prohibited the teaching of evolution in public schools, except when it was accompanied by instruction in "creation science." The Court found that, by advancing the religious belief that a supernatural being created humankind, which is embraced by the term *creation science*, the act impermissibly endorses religion. In addition, the Court found that the provision of a comprehensive science education is undermined when it is forbidden to teach evolution except when creation science is also taught. (*Edwards v. Aguillard* (1987) 482 U.S. 578)

5. In 1990, in *Webster v. New Lenox School District*, the Seventh Circuit Court of Appeals found that a school district may prohibit a teacher from teaching creation science in fulfilling its responsibility to ensure that the First Amendment's establishment clause is not violated and that religious beliefs are not injected into the public school curriculum. The court upheld a district court finding that the school district had not violated Webster's free speech rights when it prohibited him from teaching "creation science," since it is a form of religious advocacy. (*Webster v. New Lenox School District #122*, 917 F. 2d 1004)

6. In 1994, in *Peloza v. Capistrano School District*, the Ninth Circuit Court of Appeals upheld a district court finding that a teacher's First Amendment right to free exercise of religion is not violated by a school district's requirement that evolution be taught in biology classes. Rejecting plaintiff Peloza's definition of a "religion" of "evolutionism," the Court found that the district had simply and appropriately required a science teacher to teach a scientific theory in biology class. (*John E. Peloza v. Capistrano Unified School District*, (1994) 37 F. 3rd 517)

7. In 1997, in *Freiler v. Tangipahoa Parish Board of Education*, the United States District Court for the Eastern District of Louisiana rejected a policy requiring teachers to read aloud a disclaimer whenever they taught about evolution, ostensibly to promote "critical thinking." Noting that the policy singled out the theory of evolution for attention, that the only "concept" from which students were not to be "dissuaded" was "the Biblical concept of Creation," and that students were already encouraged to engage in critical thinking, the Court wrote that, "In mandating this disclaimer, the School Board is endorsing religion by disclaiming the teaching of evolution in such a manner as to convey the message that evolution is a religious viewpoint that runs counter to . . . other religious views." Besides addressing disclaimer policies, the decision is noteworthy for recognizing that curriculum proposals for "intelligent design" are equivalent to proposals for teaching "creation science." (*Freiler v Tangipahoa Board of Education*, No. 94-3577 (E.D. La. Aug. 8, 1997). On August 13, 1999, the Fifth Circuit Court of Appeals affirmed the decision; on June 19, 2000, the Supreme Court declined to hear the School Board's appeal, thus letting the lower court's decision stand.

8. In 2000, District Court Judge Bernard E. Borene dismissed the case of *Rodney LeVake v Independent School District 656, et al.* (Order Granting Defendants' Motion for Summary Judgment and Memorandum, Court File Nr. CX-99-793, District Court for the Third Judicial District of the State of Minnesota [2000]). High school biology teacher LeVake had argued for his right to teach "evidence both for and against the theory" of evolution. The school district considered the content of what he was teaching and concluded that it did not match the curriculum, which

C

The National Center
for Science Education

The National Center for Science Education, Inc. (NCSE) is a not-for-profit, membership organization that defends the teaching of evolution in the public schools. Most of its members are scientists, but many are citizens with an interest in science and education; many other members are concerned with the church and state separation issues engendered by the efforts to "balance" the teaching of evolution with the presentation of religiously based views.

And indeed, in many states and local school districts, there are ongoing efforts to eliminate or discourage the teaching of evolution, or to present religious views as science. The calls for information and assistance to NCSE steadily increase. There are now two creation science organizations funded at about $5 million per year, another major creationist organization funded at over $1 million, at least a half dozen minor ministries focusing on grassroots evangelism against evolution—and periodic antievolution assaults from television and radio evangelists listened to by millions.

Grassroots efforts require grassroots responses, and the National Center for Science Education is the only national pro-evolution organization with this grassroots focus. The NCSE recognizes that the testimony of a local science teacher or college professor at a school board meeting is more influential than a statement from a nationally prominent scientist—and the staff of the NCSE can help that local person be effective. NCSE prefers to work behind the scenes, letting our members be the "ground troops."

NCSE provides analyses of creationist arguments, background on legal and religious aspects of the controversy, and information on science education. Unfortunately, antievolutionism will not be solved by simply throwing science at it. A clear understanding of scientific aspects of the controversy is essential, of course: if creation science is scientific, then it deserves a place in the curriculum. If it is not scientific; it does not belong in the curriculum. But showing that creation science is bad science will not ensure that evolution will be taught: this requires assuaging people's fears that acceptance of evolution requires the abandonment of faith. So NCSE also must provide information beyond the scientific: connections with other citizens and groups—including religious denominations that accept evolution, and which do not want sectarian religion taught in the public schools—and advice on how to write effective letters to the editor and op-ed pieces, and on other media relations.

Calls for information to support the teaching of evolution and/or to counter antievolutionary approaches come from many directions, including state boards of education, state departments of education, committees charged with selecting textbooks, local boards of education, individual teachers, and citizens—a wide range of people and institutions. The NCSE has a speakers' bureau of knowledgeable scientists around the country who can help the public understand why evolution is an important scientific subject and why students should learn it. NCSE also maintains a Web page with information on evolution, the creation and evolution controversy, and evolution education.

The National Center for Science Education can be reached at:

National Center for Science Education
PO Box 9477
Berkeley CA 94709-0477
1-800-290-6006
ncse@ncseweb.org
www.ncseweb.org

25 Ways to Support Evolution Education

New members of NCSE often ask, "What can I do to help?" We've compiled some practical, effective suggestions from NCSE members telling us about their efforts to support evolution education. Using this list is likely to inspire you, and as you work out new ideas, be sure to share them with NCSE.

1. Donate books and videos about evolution to school and public libraries (NCSE can help you choose appropriate materials).

2. Encourage and support evolution education at museums, parks, and natural history centers (by positive remarks on comment forms, contributions to special exhibits, etc.).

3. Thank radio and television stations for including programming about evolution and other science topics.

4. Make sure friends, colleagues and neighbors know you support evolution education and can connect them with resources for promoting good science education.

5. Monitor local news media for news of anti-evolution efforts in your state or community, and inform NCSE—for example, by mailing newspaper clippings.

6. When there is controversy in your community, add your voice: Hold press conferences with colleagues, record public opinion announcements, and send letters or editorials supporting evolution education to local newspapers.

7. Ask organizations in your community to include questions about science education in questionnaires for school board candidates and other educational policy makers.

8. Share your views with school board members, legislators, textbook commissioners, and other educational policy makers.

9. Share NCSE publications with concerned citizens, educators, and colleagues.

10. Link your personal or organizational Web site to *http://www.ncseweb.org*

11. When you see a Web site that would benefit by linking to NCSE (for example, a science education site), write to the webmaster suggesting the new link (copy NCSE on the email).

12. Encourage professional and community organizations (like the PTA) to give public support to evolution education. Send copies of their public statements to NCSE.

13. Give gift subscriptions to *Reports of NCSE* to friends, colleagues, and libraries.

14. Take advantage of member benefits like discounts on book purchases and car rentals.

15. Donate to NCSE beyond your annual membership fee. (Contact NCSE about in-kind gifts and planned giving).

16. PARENTS: Make sure your child's science teacher knows s/he has your support for teaching about evolution, the age of the earth, and related concepts.

17. PARENTS: Help your child's teacher arrange field trips to natural history centers and museums with appropriate exhibits.

18. PARENTS: Discuss class activities and homework with your children—this is often the way communities learn that "creation science" is being taught; or, you may learn your child's teacher is doing a commendable job of teaching evolution.

19. PROFESSIONALS: Inform your colleagues about the evolution/creation controversy and the need for their involvement: for example, by making presentations at professional society meetings, writing articles for organizational newsletters, making announcements on email listserves.

20. COLLEGE TEACHERS: Make sure that your institution has several courses that present evolution to both majors and non-majors.

21. COLLEGE TEACHERS: Create opportunities to learn about evolution outside the classroom: for example, public lectures, museum exhibits.

22. TEACHERS: Work with your colleagues to create a supportive atmosphere in your school and community.

23. TEACHERS: Work with colleagues to develop or publicize workshops and in-service units about evolution; take advantage of them yourself.

24. INFORMAL EDUCATORS: Include evolution in signage, interpretation of exhibits, docent education, and public presentations.

25. SCIENTISTS: Share your knowledge with K-12 teachers and students by visiting classrooms or speaking at teacher-information workshops (NCSE can provide tips).

SOURCE: Eugenie Scott, Director, National Center for Science Education

D

National Association
of Biology Teachers:
Statement on Teaching Evolution

As stated in *The American Biology Teacher* by the eminent scientist Theodosius Dobzhansky (1973), "Nothing in biology makes sense except in the light of evolution." This often-quoted assertion accurately illuminates the central, unifying role of evolution in nature, and therefore in biology. Teaching biology in an effective and scientifically-honest manner requires classroom discussions and laboratory experiences on evolution.

Modern biologists constantly study, ponder and deliberate the patterns, mechanisms and pace of evolution, but they do not debate evolution's occurrence. The fossil record and the diversity of extant organisms, combined with modern techniques of molecular biology, taxonomy and geology, provide exhaustive examples and powerful evidence for genetic variation, natural selection, speciation, extinction and other well-established components of current evolutionary theory. Scientific deliberations and modifications of these components clearly demonstrate the vitality and scientific integrity of evolution and the theory that explains it.

This same examination, pondering and possible revision have firmly established evolution as an important natural process explained by valid scientific principles, and clearly differentiate and separate science from various kinds of nonscientific ways of knowing, including those with a supernatural basis such as creationism. Whether called "creation science," "scientific creationism," "intelligent-design theory," "young-earth theory" or some other synonym, creation beliefs have no place in the science classroom. Explanations employing nonnaturalistic or supernatural events, whether or not explicit reference is made to a supernatural being, are outside the realm of science and not part of a valid science curriculum. Evolutionary theory, indeed all of science, is necessarily silent on religion and neither refutes nor supports the existence of a deity or deities.

Accordingly, the National Association of Biology Teachers, an organization of science teachers, endorses the following tenets of science, evolution and biology education:

- The diversity of life on earth is the outcome of evolution: an unpredictable and natural process of temporal descent with genetic modification that is affected by natural selection, chance, historical contingencies and changing environments.

- Biological evolution refers to changes in populations, not individuals. Changes must be successfully passed on to the next generation. This means evolution results in heritable changes in a population spread over many generations. In fact, evolution can be defined as any change in the frequency of alleles within a gene pool from one generation to the next.

- Evolutionary theory is significant in biology, among other reasons, for its unifying properties and predictive features, the clear empirical testability of its integral models and the richness of new scientific research it fosters.

- The fossil record, which includes abundant transitional forms in diverse taxonomic groups, establishes extensive and comprehensive evidence for organic evolution.

- Natural selection, the primary mechanism for evolutionary changes, can be demonstrated with numerous, convincing examples, both extant and extinct.

- Natural selection—a differential, greater survival and reproduction of some genetic variants within a population under an existing environmental state—has no specific direction or goal, including survival of a species.

- Adaptations do not always provide an obvious selective advantage. Furthermore, there is no indication that adaptations—molecular to organismal—must be perfect: adaptations providing a selective advantage must simply be good enough for survival and increased reproductive fitness.

- The model of punctuated equilibrium provides another account of the tempo of speciation in the fossil record of many lineages: it does not refute or overturn evolutionary theory, but instead adds to its scientific richness.

- Evolution does not violate the second law of thermodynamics: producing order from disorder is possible with the addition of energy, such as from the sun.

- Although comprehending deep time is difficult, the earth is about 4.5 billion years old. *Homo sapiens* has occupied only a minuscule moment of that immense duration of time.

- When compared with earlier periods, the Cambrian explosion evident in the fossil record reflects at least three phenomena: the evolution of animals with readily-fossilized hard body parts; Cambrian environment (sedimentary rock) more conducive to preserving fossils; and the evolution from pre-Cambrian forms of an increased diversity of body patterns in animals.

- Radiometric and other dating techniques, when used properly, are highly accurate means of establishing dates in the history of the planet and in the history of life.

- Recent findings from the advancing field of molecular genetics, combined with the large body of evidence from other disciplines, collectively provide indisputable demonstration of the theory of evolution.

- In science, a theory is not a guess or an approximation but an extensive explanation developed from well-documented, reproducible sets of experimentally-derived data from repeated observations of natural processes.

- The models and the subsequent outcomes of a scientific theory are not decided in advance, but can be, and often are, modified and improved as new empirical evidence is uncovered. Thus, science is a constantly self-correcting endeavor to understand nature and natural phenomena.

- Science is not teleological: the accepted processes do not start with a conclusion, then refuse to change it, or acknowledge as valid only those data that support an unyielding conclusion. Science does not base theories on an untestable collection of dogmatic proposals. Instead, the processes of science are characterized by asking questions, proposing hypotheses, and designing empirical models and conceptual frameworks for research about natural events.

- Providing a rational, coherent and scientific account of the taxonomic history and diversity of organisms requires inclusion of the mechanisms and principles of evolution.

- Similarly, effective teaching of cellular and molecular biology requires inclusion of evolution.

- Specific textbook chapters on evolution should be included in biology curricula, and evolution should be a recurrent theme throughout biology textbooks and courses.

- Students can maintain their religious beliefs and learn the scientific foundations of evolution.

- Teachers should respect diverse beliefs, but contrasting science with religion, such as belief in creationism, is not a role of science. Science teachers can, and often do, hold devout religious beliefs, accept evolution as a valid scientific theory, and teach the theory's mechanisms and principles.

- Science and religion differ in significant ways that make it inappropriate to teach any of the different religious beliefs in the science classroom.

Opposition to teaching evolution reflects confusion about the nature and processes of science. Teachers can, and should, stand firm and teach good science with the acknowledged support of the courts. In *Epperson v. Arkansas* (1968), the U.S. Supreme Court struck down a 1928 Arkansas law prohibiting the teaching of evolution in state schools. In *McLean v.*

Arkansas (1982), the federal district court invalidated a state statute requiring equal classroom time for evolution and creationism.

Edwards v. Aguillard (1987) led to another Supreme Court ruling against so-called "balanced treatment" of creation science and evolution in public schools. In this landmark case, the Court called the Louisiana equal-time statute "facially invalid as violative of the Establishment Clause of the First Amendment, because it lacks a clear secular purpose." This decision—"the Edwards restriction"—is now the controlling legal position on attempts to mandate the teaching of creationism: the nation's highest court has said that such mandates are unconstitutional. Subsequent district court decisions in Illinois and California have applied "the Edwards restriction" to teachers who advocate creation science, and to the right of a district to prohibit an individual teacher from promoting creation science, in the classroom.

Courts have thus restricted school districts from requiring creation science in the science curriculum and have restricted individual instructors from teaching it. All teachers and administrators should be mindful of these court cases, remembering that the law, science and NABT support them as they appropriately include the teaching of evolution in the science curriculum.

References and Suggested Reading

Aguillard, D. (1999). Evolution education in Louisiana public schools: a decade following *Edwards v. Aguillard*. *The American Biology Teacher, 61*, pp. 182–188.

Brack, A. (Ed.) (1999). *The Molecular Origins of Life: Assembling Pieces of the Puzzle*. Cambridge: Cambridge University Press.

Futuyma, D. (1986). *Evolutionary Biology*, 2nd ed. Sunderland, MA: Sinauer Associates, Inc.

Futuyma, D. (1995). *Science on Trial*. Sunderland, MA: Sinauer Associates, Inc.

Gillis, A. (1994). Keeping creationism out of the classroom. *Bioscience, 44*, pp. 650–656.

Gould, S. (1994, October). The evolution of life on earth. *Scientific American, 271*, pp. 85–91.

Gould, S. (1995). *Dinosaur in a Haystack. Reflections in Natural History*. New York: Harmony Books.

Kiklas, K. (1997). *The Evolutionary Biology of Plants*. Chicago: The University of Chicago Press.

Matsumura, M. (Ed.). (1995). *Voices for Evolution*. Berkeley, CA: The National Center for Science Education.

Mayr, E. (1991). *One Long Argument: Charles Darwin and the Genesis of Modern Evolutionary Thought*. Cambridge, MA: Harvard University Press.

Moore, J. (1993). *Science as a Way of Knowing—The Foundations of Modern Biology*. Cambridge, MA: Harvard University Press.

Moore, R. (1999). Creationism in the United States: VII. The Lingering Threat. *The American Biology Teacher, 61*, pp. 330–340. See also references therein to earlier articles in the series.

National Academy of Sciences. (1998). *Teaching About Evolution and the Nature of Science.* Washington, DC: National Academy Press.

National Academy of Sciences. (1999). *Science and creationism—A View from the National Academy of Sciences.* Washington, DC: National Academy Press.

National Center for Science Education. P.O. Box 9477, Berkeley, CA 94709. Numerous publications such as Bartelt, K. (1999), *A Scientist Responds to Behe's Black Box.*

National Research Council. (1996). *National Science Education Standards.* Washington, DC: National Academy Press.

Pennock, R.T. (1999). *Tower of Babel: The Evidence Against the New Creationism.* Cambridge, MA: MIT Press.

Weiner, J. (1994). *Beak of the Finch—A Story of Evolution in our Time.* New York: Alfred A. Knopf.

Wilson, E. (1992). *The Diversity of Life.* New York: W.W. Norton & Co.

Adopted by the Board of Directors March 15, 1995. Revised October 1997 and August 2000. Endorsed by: The Society for the Study of Evolution, June 1998. The American Association of Physical Anthropologists, July 1998.

E

An NSTA Position Statement: The Teaching of Evolution

Introductory Remarks

The National Science Teachers Association supports the position that evolution is a major unifying concept of science and should be included as part of K–College science frameworks and curricula. NSTA recognizes that evolution has not been emphasized in science curricula in a manner commensurate to its importance because of official policies, intimidation of science teachers, the general public's misunderstanding of evolutionary theory, and a century of controversy.

Furthermore, teachers are being pressured to introduce creationism, creation "science," and other nonscientific views, which are intended to weaken or eliminate the teaching of evolution.

Within this context, NSTA recommends that:

- Science curricula and teachers should emphasize evolution in a manner commensurate with its importance as a unifying concept in science, and its overall explanatory power.

- Policy makers and administrators should not mandate policies requiring the teaching of creation science, or related concepts such as so-called "intelligent design," "abrupt appearance," and "arguments against evolution."

- Science teachers should not advocate any religious view about creation, nor advocate the converse: that there is no possibility of supernatural influence in bringing about the universe as we know it. Teachers should be nonjudgmental about the personal beliefs of students.

- Administrators should provide support to teachers as they design and implement curricula that emphasize evolution. This should include inservice education to assist teachers to teach evolution in a comprehensive and professional manner. Administrators also should support teachers against pressure to promote nonscientific views or to diminish or eliminate the study of evolution.

- Parental and community involvement in establishing the goals of science education and the curriculum development process should be encouraged and nurtured in our democratic society. However, the professional responsibility of science teachers and curriculum specialists to provide students with quality science education should not be bound by censorship, pseudoscience, inconsistencies, faulty scholarship, or unconstitutional mandates.

- Science textbooks shall emphasize evolution as a unifying concept. Publishers should not be required or volunteer to include disclaimers in textbooks concerning the nature and study of evolution.

NSTA offers the following background information:

The Nature of Science, and Scientific Theories

Science is a method of explaining the natural world. It assumes the universe operates according to regularities and that through systematic investigation we can understand these regularities. The methodology of science emphasizes the logical testing of alternate explanations of natural phenomena against empirical data. Because science is limited to explaining the natural world by means of natural processes, it cannot use supernatural causation in its explanations. Similarly, science is precluded from making statements about supernatural forces, because these are outside its provenance. Science has increased our knowledge because of this insistence on the search for natural causes.

The most important scientific explanations are called "theories." In ordinary speech, "theory" is often used to mean "guess," or "hunch," whereas in scientific terminology, a theory is a set of universal statements that explain the natural world. Theories are powerful tools. Scientists seek to develop theories that

- are internally consistent and compatible with the evidence

- are firmly grounded in and based upon evidence

- have been tested against a diverse range of phenomena

- possess broad and demonstrable effectiveness in problem-solving

- explain a wide variety of phenomena.

The body of scientific knowledge changes as new observations and discoveries are made. Theories and other explanations change. New theories emerge and other theories are modified or discarded. Throughout this process, theories are formulated and tested on the basis of evidence, internal consistency, and their explanatory power.

Evolution as a Unifying Concept

Evolution in the broadest sense can be defined as the idea that the universe has a history: that change through time has taken place. If we look today at the galaxies, stars, the planet earth, and the life on planet earth, we see that things today are different from what they were in the past: galaxies, stars, planets, and life forms have evolved. Biological evolution refers to the scientific theory that living things share ancestors from which they have diverged: Darwin called it "descent with modification." There is abundant and consistent evidence from astronomy, physics, biochemistry, geochronology, geology, biology, anthropology and other sciences that evolution has taken place.

As such, evolution is a unifying concept for science. The National Science Education Standards recognizes that conceptual schemes such as evolution "unify science disciplines and provide students with powerful ideas to help them understand the natural world," and recommends evolution as one such scheme. In addition, the Benchmarks for Science Literacy from the American Association for the Advancement of Science's Project 2061, and the NSTA's Scope, Sequence, and Coordination Project as well as other national calls for science reform, all name evolution as a unifying concept because of its importance across the discipline of science. Scientific disciplines with a historical component such as astronomy, geology, biology, and anthropology, cannot be taught with integrity if evolution is not emphasized.

There is no longer a debate among scientists over whether evolution has taken place. There is considerable debate about how evolution has taken place: the processes and mechanisms producing change, and what has happened during the history of the universe. Scientists often disagree about their explanations. In any science, disagreements are subject to rules of evaluation. Errors and false conclusions are confronted by experiment and observation, and evolution, as in any aspect of science, is continually open to and subject to experimentation and questioning.

Creationism

The word "creationism" has many meanings. In its broadest meaning, creationism is the idea that a supernatural power or powers created. Thus to Christians, Jews, and Muslims, God created; to the Navajo, the Hero Twins created. In a narrower sense, "creationism" has come to mean "special creation": the doctrine that the universe and all that is in it was created by God in essentially its present form, at one time. The most common variety of special creationism asserts that

- the earth is very young

- life was originated by a creator

- life appeared suddenly

- kinds of organisms have not changed

- all life was designed for certain functions and purposes.

This version of special creation is derived from a literal interpretation of Biblical Genesis. It is a specific, sectarian religious belief that is not held by all religious people. Many Christians and Jews believe that God created through the process of evolution. Pope John Paul II, for example, issued a statement in 1996 that reiterated the Catholic position that God created, but that the scientific evidence for evolution is strong.

"Creation science" is an effort to support special creationism through methods of science. Teachers are often pressured to include it or synonyms such as "intelligent design theory," "abrupt appearance theory," "initial complexity theory," or "arguments against evolution" when they teach evolution. Special creationist claims have been discredited by the available evidence. They have no power to explain the natural world and its diverse phenomena. Instead, creationists seek out supposed anomalies among many existing theories and accepted facts. Furthermore, creation science claims do not provide a basis for solving old or new problems or for acquiring new information.

Nevertheless, as noted in the National Science Education Standards "Explanations on how the natural world changed based on myths, personal beliefs, religious values, mystical inspiration, superstition, or authority may be personally useful and socially relevant, but they are not scientific." Because science can only use natural explanations and not supernatural ones, science teachers should not advocate any religious view about creation, nor advocate the converse: that there is no possibility of supernatural influence in bringing about the universe as we know it.

Legal Issues

Several judicial rulings have clarified issues surrounding the teaching of evolution and the imposition of mandates that creation science be taught when evolution is taught. The First Amendment of the Constitution requires that public institutions such as schools be religiously neutral; because special creation is a specific, sectarian religious view, it cannot be advocated as "true," accurate scholarship in the public schools. When Arkansas passed a law requiring "equal time" for creationism and evolution, the law was challenged in Federal District Court. Opponents of the bill included the religious leaders of the United Methodist, Episcopalian, Roman Catholic, African Methodist Episcopal, Presbyterian, and Southern Baptist churches, and several educational organizations. After a full trial, the judge ruled that creation science did not qualify as a scientific theory (*McLean v. Arkansas Board of Education,* 529 F. Supp. 1255 (ED Ark. 1982)).

Louisiana's equal time law was challenged in court, and eventually reached the Supreme Court. In *Edwards v. Aguillard,* 482 U.S. 578 (1987) the court determined that creationism was inherently a religious idea and to mandate or advocate it in the public schools would be unconstitutional. Other court decisions have upheld the right of a district to require that a teacher teach evolution and not to teach creation science (*Webster v. New Lennox School District #122,* 917 F.2d 1003 (7th Cir. 1990); *Peloza v. Capistrano Unified School District,* 37 F.3d 517 (9th Cir. 1994)).

Some legislatures and policy makers continue attempts to distort the teaching of evolution through mandates that would require teachers to teach evolution as "only a theory," or that require a textbook or lesson on evolution to be preceded by a disclaimer. Regardless of the legal status of these mandates, they are bad educational policy. Such policies have the effect of intimidating teachers, which may result in the de-emphasis or omission of evolution. The public will only be further confused about the special nature of scientific theories, and if less evolution is learned by students, science literacy itself will suffer.

—Adopted by the NSTA Board of Directors in July, 1997

References

Aldridge, Bill G. (Ed.). (1996). *Scope, Sequence, and Coordination: A High School Framework for Science Education.* Arlington, VA: National Science Teachers Association (NSTA).

American Association for the Advancement of Science (AAAS), Project 2061. (1993). *Benchmarks for Science Literacy.* New York: Oxford University Press.

Daniel v. Waters, 515 F.2d 485 (6th Cir. 1975).

Edwards v. Aguillard, 482 U.S. 578 (1987).

Epperson v. Arkansas, 393 U.S. 97 (1968).

Laudan, Larry. (1996). *Beyond Positivism and Relativism: Theory, Method, and Evidence.* Boulder, CO: Westview Press.

McLean v. Arkansas Board of Education, 529 F. Supp. 1255 (ED Ark. 1982).

National Research Council. (1996). *The National Science Education Standards.* Washington, DC: National Academy Press.

National Science Teachers Association (NSTA). (1993). *Scope, Sequence, and Coordination of Secondary School Science. Vol. 1. The Content Core: A Guide for Curriculum Designers (Rev. ed).* Arlington, VA: Author.

Peloza v. Capistrano Unified School District, 37 F.3d 517 (9th Cir. 1994).

Ruse, Michael. (1996). *But Is It Science: The Philosophical Question in the Creation/Evolution Controversy.* Amherst, NY: Prometheus.

Webster v. New Lennox School District #122, 917 F.2d 1003 (7th Cir. 1990).

F

Evolution Education Research Centre

Having opened its doors in 2000, the Evolution Education Research Centre (EERC) is an academic body with a mission to advance the teaching and learning of biological evolution through research. The center's research focus is on investigating ways to increase evolution understanding, with the goal of improving the teaching and learning of evolution at all educational levels.

EERC's international team currently consists of four research professors from McGill University, Montreal, and four from Harvard University, Cambridge, who have combined expertise in anthropology, biological evolution, educational psychology, evolution education, geology, molecular biology, paleontology, philosophy of science, philosophy of education, and science education. The overarching research objectives of EERC are to design and conduct studies that will inform the practice of evolution education.

Address:

Brian J. Alters, Director
Evolution Education Research Centre
McGill University
Faculty of Education
3700 McTavish Street
Montreal, QC H3A 1Y2 Canada

Notes

Introduction

1 (People for the American Way Foundation 2000, p. 4)
2 (People for the American Way Foundation 2000, p. 45)
3 (See Chapter 2 for a more detailed characterization of creationism)
4 For example, polling conducted by Gallup news service (Gallup 1999) and the (National Science Board 2000, p. A-549).
5 (Gallup 1999)
6 (Gillespie 1999)
7 (National Science Board 2000, p. A-549)
8 (People for the American Way Foundation 2000, p. 38)
9 (People for the American Way Foundation 2000, p. 41)
10 (People for the American Way Foundation 2000, p. 40)
11 (People for the American Way Foundation 2000, p. 42)
12 (Hemenway 1999)
13 J. Morris 2000, July. Open letter to ICR mailing list; (Beem 2000).
14 Moore quoted in (Christensen 1998).
15 (Moore, R. 1999)
16 (People for the American Way Foundation 2000, p. 5)
17 (Futuyma 2000)

Chapter 1

1 (Ham 1987, sixteenth printing 1998, p. 97)
2 (Futuyma 2000, p. 43) The nine scientific societies endorsing the document are the American Society of Naturalists, American Behavior Society, Ecological Society of America, Genetics Society of America, Paleontological Society, Society for Molecular Biology and Evolution, Society for the Study of Evolution, Society of Systematic Biologists, and American Institute of Biological Sciences.
3 (Moore, R. 1999, p. 336)
4 (Morris and Morris 1996, Vol. 1, p. 167)
5 (Morris, J. 1990)
6 (Johnson 1997, p. 9)
7 (Johnson 1993, p. 146)
8 (Marks 1995, back cover)
9 (Marks 1995, pp. 100–101)
10 (Answers in Genesis, no date[a])
11 (Gallup and Newport 1991)
12 New American Standard Bible
13 (McDowell 1993, pp. 394–395)
14 (McDowell and Stewart 1980, pp. 100–101)
15 (Institute for Creation Research 1999)
16 (InterVarsity Christian Fellowship/USA, 1999, 2000)
17 (InterVarsity Press 1999, back inside cover)
18 (InterVarsity Press 1999, p. 35)

19   Even though many in Behe's camp contend he is not a creationist, his writings have led both evolutionists and creationists to a creationist characterization. He seems to accept the evidence for common ancestry of humans with the great apes (Miller 1999, p. 164) but conversely feels that Darwin's theory has failed on the molecular scale and, therefore, that the scientific community should conclude intelligent causality. Behe writes: "The impotence of Darwinian theory in accounting for the molecular basis of life is evident. . . . But we are here. . . . Clearly, if something was not put together gradually, then it must have been put together quickly or even suddenly." He ends his book, *Darwin's Black Box*, by concluding "that life was designed by an intelligent agent" (Behe 1996, pp. 187 & 252). The most prolific creationist, Henry Morris, considers Behe to be a "Catholic creationist" (Morris, H. 1998b), and philosopher of science Robert Pennock, who recently authored an extensive book on the newest type of creationists, considers Behe's views a variation of creationism (Pennock 1999). Furthermore, in a highly publicized television debate on *The Firing Line with William F. Buckley,* Michael Ruse called Behe a creationist to his face (aired 12/19/97 PBS). Ruse stated: "Unlike so many creationists, certainly creationists in the past, you haven't just attacked other people or other people's theories." Behe did not rebut the creationist label.

20   (InterVarsity Press 1999, p. 22)

21   (Focus on the Family 1998a)

22   (Focus on the Family 1998b)

23   (Kaufman 1998)

24   (Wehmeyer 2000)

25   (Focus on the Family 2000)

26   (Focus on the Family 1998c)

27   (National Academy of Sciences 1998)

28   (Hartwig 1998)

29   (Kinney 2000; Van Campen 1999; Answers in Genesis 2001; Ham 2001, p. 7)

30   (Answers in Genesis, no date[b])

31   (Institute for Creation Research, no date[b])

32   (Morris, J. 1999, February)

33   (Institute for Creation Research 2000b, pp. 1–2)

34   (Institute for Creation Research 2000d, p. 3; Institute for Creation Research 2000e; Morris, H. 1995b)

35   (Ackerman and Williams 1999, p. 62)

36   (Discovery Institute: Center for the Renewal of Science & Culture, no date)

37   (Numbers 1998, p. 11)

38   (Association de Science Créationniste du Québec)

39   Numbers 2000, personal communication; (Institute for Creation Research 2000e)

40   (Morris, J. 1999, March)

41   (Gallup News Service 2000)

42   (Morris, J. 1990)

43   (Morris, H. 1995a, Introduction)

44   We are not saying that the creationists originated, or are the only ones using, militaristic symbolism in discussions concerning evolution/creation. In fact, Adrian Desmond, the great historian of nineteenth-century British science, recounts that it is actually Thomas Henry Huxley (Darwin's Bulldog, 1825–1895) that encouraged much of the military semantics. "We owe him that enduring military metaphor, the 'war' of science against theology." (Desmond 1997, p. xiii). We simply are attempting to illustrate that, while many science instructors do not

consider themselves in any war, symbolic or not, many current creationists have included them in their war, like it or not.

45 (Ham 1998, p. 81)
46 (Morris, J. 1999, p. 88)
47 (Morris, H. 1997a)
48 (Morris, J. 1999)
49 (Wells 2000; Morris, H. 1998a)
50 (Numbers 1998)
51 (Morris, H. 1998b)
52 (Morris, H. 1997a)
53 (Morris, J. 1990)
54 There are many prominent evolutionary biologists who possess a worldview based on evolutionary humanism with an understanding that even an ethical system could and possibly should be based upon it. Among popular advocates of such positions are renowned scientists such as Ernst Mayr, Richard Dawkins, and E. O. Wilson. Ernst Mayr writes in the tenth decade of his life about such a worldview under the heading of "What Moral System Is Best Suited for Humankind?"

> The traditional ethical norms of Western culture are those of the Judeo-Christian tradition, that is, they are based on the various commandments and injunctions articulated in the Old and New Testaments. . . . The traditional norms of the West are no longer adequate. . . . Evolution does not provide us with a codified set of ethical norms such as the Ten Commandments. Yet evolution did give us a capacity for stretching beyond our individual needs to take those of the larger group into account. And an understanding of evolution can give us a worldview that serves as the basis for a sound ethical system. (Mayr 1997, pp. 265–270)

> Is there any particular ethics that an evolutionist should adopt? Ethics is a very private matter, a personal choice. My own values are rather close to Julian Huxley's evolutionary humanism. "It is a belief in mankind, a feeling of solidarity with mankind, and a loyalty toward mankind. Man is the result of millions of years of evolution, and our most basic ethical principle should be to do everything toward enhancing the future of mankind. All other ethical norms can be derived from this baseline." (Mayr 1997, p. 269)

55 (Larson and Witham 1997)
56 (Morris and Morris 1996, Vol. 1, p. 181)
57 (Morris and Morris 1996, Vol. 1, p. 178)
58 (Morris and Morris 1996, Vol. 1, pp. 183-184)
59 (Deckard 1998, p. iii)
60 (Morris, J. and Phillips 1998)
61 (Marks 1995)
62 (Morris and Morris 1996, Vol. 1, pp. 181–182)
63 (Gould 1999, p. 149)

Chapter 2

1 (Gish 1995b, 63)
2 (Ham 1987, pp. 97–98) Sixteenth printing 1998.
3 (Numbers 1998)
4 (Gallup 1999)

5   (McCourt 1998)

6   Quoted by (Maudlin 1998, p. 4)

7   (Wilson 1997, p. 11)

8   (Whitcomb and Morris, 1961)

9   (Now is the Time 1997, p. 27)

10 We do not consider all theists to be creationists. Some theists believe that God created by using evolution, and that His involvement in evolution cannot be detected by science. Further, they feel that these beliefs should not be taught as science. Therefore, by the definition of creationists we use for the purposes of this book (given in the Introduction), such advocates are not considered creationists.

11 (People for the American Way Foundation 2000, p. 33)

12 (Institute for Creation Research, no date (c))

13 (Reasons to Believe, no date)

14 (Audi 1995, p. 167)

15 (McInerney 1997)

16 See (Van Till 1996)

17 For example (Gould 1989)

18 (Morris, J. 1999, August. Open letter to ICR mailing list)

19 Many instructors have told us that they not only read public opinion polls but also give the questions to their students to find out their views on evolution prior to instruction. We agree that accessing students' prior knowledge about evolution is important to good instruction, but we find that many of the public opinion polls do not have sufficient specificity for learning what students really think. Of course, many science educators believe that in small classes, open-ended questions serve best. However, the advantages of this position must be balanced against some students' intentional lack of giving full answers due to their feeling that the instructor may be able to recognize or match their handwriting to their later work. In other words, when it comes accessing students' views about what many believe to be the most sensitive topic in the course, rather than using an open-ended questionnaire, a well designed anonymous multiple choice questionnaire may more accurately reflect students' real views. Another advantage of such a self-designed questionnaire is the omission of questions that most would consider overtly religious. For example, if you wanted to learn more about what students really thought about the age of the earth, instead of using something like "God created the earth in the last 10,000 years," a better approach would be to use "The earth is approximately 6,000 years old," "The earth is approximately 10,000 years old," "The earth is approximately 100,000 years old," "The earth is approximately a million years old," and "The earth is approximately a billion years old." Each item would be followed by a Likert scale (e.g., strongly agree, agree, undecided, disagree, strongly disagree).

20 (Gallup 1999)

21 As stated previously, for the purposes of this book we do not consider these individuals to be creationists because their views do not fit our working definition for creationism as described in the Introduction.

22 (People for the American Way Foundation 2000)

23 (Ross 1994, p. 83)

24 (Institute for Creation Research, no date (c))

25 (Pinker 1997a, pp. 360–361)

26 (Institute for Creation Research, no date (c))

27 (Pennock 1999; King 2000)

28 See p. 122 for a rationale why the Raëlian view is not considered science.
29 (People for the American Way Foundation 2000)
30 (Johnson 1995, p. 208)
31 (Johnson 1995, pp. 37–38)
32 However, some have written about how science and religion can coexist with mutual respect and noninterference; for example (Gould, 1999).

NOTE: Parts of this chapter are adapted from (Alters 1999b).

Chapter 3

1 (Darwin 1898, Vol. II, p. 294)
2 In this chapter, when we do not specify the type of creationism, we usually default to literalism.
3 We use mostly the King James version because that is the version typically cited by literalist creationist leaders.
4 (Morris and Morris 1996, Vol. 1)
5 (Ryrie 1994, p. vi)
6 (Morris, H. 1995a, cover)
7 (Morris, H. 1995a, p. 1015)
8 (Ryrie 1994, p. 4)
9 (Morris, H. 1995a, p. 4)
10 (MacArthur 1997, p. 16)
11 (Shelley 1994, p. 1)
12 (Shelley 1994, p. 11)
13 (Shelley 1994, p. 109)
14 (Morris, H. 1993, p. 14)
15 (Morris, H. 1993, pp. 271–272)
16 (Morris, J. 1996, p. a)
17 (Institute for Creation Research, no date[a])
18 (Institute for Creation Research, no date[c])
19 Of course there are institutions that are not neutral in their teaching of the subject. For example, Liberty University, "a Christian liberal arts university with over 5,000 resident students as well as several thousand students in its external degree program" had a telling advertisement in the *Chronicle of Higher Education* for the 1999–2000 academic year: "Assistant Professor of Biology. Expertise in some area of creation/evolution essential. Experience with molecular techniques helpful." (Chronicle of Higher Education Career Network 1998)
20 For example, Biola University in California (Biola University, no date).
21 (Morris, H. 1989, p. 208)

NOTE: Parts of this chapter are adapted from (Alters 1999a).

Chapter 4

1 (Morris and Morris 1996, Vol. 2, p. 107)
2 (Eldridge 2000, p. 14)
3 (Wandersee, Mintzes, and Novak 1994, p. 199).
4 (Kitcher 1982) Eighth printing 1994.
5 (Laudan 1983, p. 347)
6 (Kitcher 1982, p. 44)

7   (Mayr 1997, p. 49)

8   For example (Hung 1997 and McGuire 1992)

9   For further explanation see (Mayr 1997).

10   (Morris and Morris 1996, Vol. 2)

11   (National Academy of Sciences 1999)

12   (Morris, H. 1995c; 1976)

13   (Gish 1995A , p. 51)

14   (Carroll, 1988)

15   (Morris, H. 1997b)

16   In addition to using out-of-context quotes, creationists have been known to "distort the words and works of scientists." For one victim's accounts of such behavior, see (Eldredge 2000, p. 129).

17   A useful family tree is contained in the journal *Natural History* (Shoshani 1997, p. 38).

18   Ken Miller of Brown University has used this approach quite successfully in formal debate with leading creationists and has a reproduction of the proboscidean lineage in his book (Miller 1999). Also see (Eldredge 2000).

19   (Eldredge 2000)

20   (Eldredge and Gould 1972)

21   (Gish, 1974, p. 20)

22   (Morris, J. 1976, p. ii)

23   (Institute for Creation Research 1981, p. 1)

24   (Morris, J. 1980, p. 4)

25   (Godfrey and Cole 1986)

26   (Morris, J. 1996, p. c)

27   (Cole 1995, p. 1)

28   (Gish 1986, Revised 1994, p. 20)

29   (Dobzhansky 1967)

30   (Behe 1996, p. 252)

31   (Behe 1996, p. 5)

32   (Morris and Morris 1996, Vol. 2, p. 175)

33   For good review of origin of life science, see (Strickberger 2000).

34   For a collection of such statements, see (Matsumura 1995).

35   (Morris and Morris 1996, Vol. 2, p. 270)

Chapter 5

1   (Christensen 1998)

2   (Gardner 1999a, p. 16)

3   (World Health Organization 2000)

4   (World Health Organization 2000)

5   (Pennisi 2000)

6   (Futuyma 1995, p. 48)

7   (National Research Council 1996)

8   (American Association for the Advancement of Science 1993)

9   (National Science Teachers Association 1997) and in Appendix E, p. 219

10   NABT Statement on Teaching Evolution. Available: http://www.nabt.org/evolution.html and in Appendix D, p. 213

11   (Futuyma, D. 2000)

Chapter 6

1   (Numbers 1998, p. 20)

2   We understand that many of the questions we selected to address in Chapters 6–8 can be considered by some to be categorized better in chapters other than the ones we used. For example, many people would consider the question "Are you telling me that miracles don't happen?" a religious question; however, some would place it in the science category because the majority of our answer is nonreligious. This is a matter of one's perspective on categorizing responses that sometimes appear to overlap. We have attempted to categorize questions by where we think the majority of science instructors might look for them.

3   However, as discussed in greater detail in Chapter 2, many creationists would consider "these" Christians to be leading nonexemplary Christian lives due, in part, to their acceptance of what they consider to be the false doctrine of evolution, and due, in part, to the sinful nature of all mankind.

4   (Morris and Morris 1996, Vol. 2, p. 14)

5   (Scott 1995, p. 10)

6   Ernst Mayr writes that "There are many branches of science in which prediction plays a very subordinate role." "For the biologist, it is not so important that his theory survive the test of prediction; it is more important that his theory is useful in solving problems" (Mayr 1997, pp. 25 & 54).

7   (Gould and Lewontin 1979)

8   (Miller 1999, p. 102)

9   Evolutionary biologist E. O. Wilson estimates that there may be as many as 100 million living species on the planet (Wilson 1992, 2000, p. 132), and paleontologist George Gaylord Simpson estimated roughly that, in 3.5 billion years of life on earth, as many as 50 billion species may have existed (Simpson 1952). Also see statistical paleontologist Raup (Raup 1991).

10 For one of the most famous works on biological imperfections in living organisms, see Stephen Jay Gould's book *The Panda's Thumb* (Gould, 1980). For examples of detailed explanations of how biological complexities—example, the human eye—could arise by evolutionary mechanisms, see works by Richard Dawkins (Dawkins 1987, 1996).

11 For a detailed rebuttal of creationists' intelligent design SETI arguments, see Robert Pennock's *Tower of Babel: The Evidence against the New Creationism* (Pennock 1999, Chapter 5).

12 (Gilchrist 1997, p. 15) At this writing, George Gilchrist was at Clarkson University in Potsdam, NY. He conducted the research cited here while at University of Washington.

13 Forrest 2000, Personal communication. At the time of this writing, Forrest was preparing her study for publication.

14 (Gingerich 1992, p. 254)

15 (Bell 1997, p. 62)

16 (Wheeler 1996, pp. A13, A16)

17 (Alters 1996, pp. B3–B12)

18 (Mayr 1997, p. 61)

19 (Mayr 1997, p. 26)

20 (Mayr 1997, p. 61)

21 (Morris and Morris 1996, Vol. 2, p. 70)

22 (Gould and Lewontin 1979; Pinker 1997b)

23 (Gould 1977, p. 14)

24 (Lewontin 2000)

25   Creationists commonly disagree with many scientific hypotheses that help confirm the occurrence of evolution. Some of these hypotheses are less well known among the general public, and others are portions of major well-known, established scientific theories that support evolution. (For example, the literalists have considerable problems with plate tectonics because of the time element.)

26   (Huxley 1882, p. 312)

27   (Mayr 1997, p. 179)

28   (Johnson 1997, pp. 114–115)

29   Given that Johnson is in the field of law, not in science or science education, we chose an example that, when applied in the field of law, would be as illogical and unacceptable to the profession as it would be in science and science education.

Chapter 7

1   (Morris, H. 1997b, p. iii)

2   (Desmond and Moore 1991, p. 636)

3   (Desmond and Moore 1991, p. 658)

4   (Morris, H. 1989, p. 95)

5   (Desmond 1997, p. 631)

6   (Moore 1994, p. 111)

7   (Moore 1994, p. 100)

8   (Morris, H. 1989, p. 95)

9   (Numbers 1992)

10   (Morris, H. and Parker 1982, p. 264)

11   (U. S. Supreme Court Education Cases 1991, pp. 91-92)

12   (Morris, H. 1995a, p. 1015)

13   (McDowell 1993, p. 395)

14   (Moreland 1989, p. 226)

Chapter 8

1   (Gould 1999, p. 125)

2   (Overton 1982, p. 318)

3   (Lawson and Worsnop 1992)

4   (Gardiner 1998, p. 73)

5   Ernst Mayr is even more liberal concerning how biological theories are violated: "What has been increasingly appreciated is that the assessment of a theory is not a matter of simple logical rules and that rationality has to be construed in broader terms than either deductive or inductive logic offer" (Mayr 1997, p. 51).

6   (Eldredge 2000)

7   (Gould 1978, p. 509)

Chapter 9

1   (Kennedy 1998)

2   (Scott 1997, p. 285)

3   (Institute for Creation Research 2000c, p. 10)

4   (Ham 1998, p. 73)

5   (Morris, J. and Phillips 1998, p. 5)

6   (Matthews, M. 1996)
7   (Overman and Deckard 1997)
8   (Moore 1999, p. 331)
9   (Ehrenreich 1999, p. 45)
10 (Scott 1998, p. 25)

Chapter 10

1   (Gardner 1999a, p. 16)
2   (Tobin, Tippins, and Gallard 1994, p. 48)
3   (Sadler 1998)
4   (Harrison, Grayson, and Treagust 1999)
5   (Settlage 1994)
6   (Lawson 1994, p. 166)
7   (Rea-Ramirez and Clement 1998; Beeth 1995; Duschl and Gitomer 1991; Posner, Strike, Hewson, and Gertzog 1982)
8   (Bishop and Anderson 1990, p. 418)
9   (Greene 1990, p. 878)
10 (Bishop and Anderson 1986)
11 (Brumby 1984)
12 (Bishop and Anderson 1990; Bishop and Anderson 1986)
13 (Brumby 1984)
14 (Vincenzo Bizzo 1994)
15 (Bishop and Anderson 1990)
16 (Osborne and Wittrock 1983)
17 (Brumby 1984, p. 500)
18 (Brumby 1984, p. 500)
19 (Beeth 1995, p. 4)
20 (Jimenez Aleixandre 1994)
21 (Brumby 1984)
22 (Osborne and Wittrock 1983)
23 (Scott 1992)
24 (Osborne and Wittrock 1983)
25 (Brumby 1984; Posner, Strike, Hewson, and Gertzog 1982)
26 (Roth, Anderson, and Smith 1987) Even though it overtly deals with fifth-grade students, this article contains useful information about questioning and responding to any level of student to foster conceptual change.
27 (Bishop and Anderson 1990, p. 422)
28 (Jensen and Finley 1995; Jensen and Finley 1996; Jensen and Finley 1997)
29 (Wandersee 1985; Clough and Wood-Robinson 1985; Lawson and Weser 1990)
30 (American Association for the Advancement of Science 1993, pp. 237–260)
31 (Lawson and Weser 1990, p. 605)
32 (Jensen and Finley 1997)
33 (Jensen and Finley 1997, p. 211)
34 (Zuzovsky 1994)
35 (Bishop and Anderson 1986)
36 (Bishop and Anderson 1990; Demastes, Settlage, and Good 1995)

37 (National Research Council 1996, p. 214)

38 (BSCS 1992)

39 (Settlage 1994; Demastes, Settlage, and Good 1995)

40 (National Research Council 1996, pp. 182–183)

41 (Wandersee 1990)

42 (Wandersee 1990, p. 927)

43 (Wandersee 1990, p. 933)

44 (Trowbridge and Wandersee 1994)

45 (Wandersee, Mintzes, and Novak 1994)

46 (Lawson, Abraham, and Renner 1989)

47 (Trowbridge, Bybee, and Carlson Powell 2000)

48 (Scharmann 1993, first two examples; O'Brien 2000, third example)

49 (Keown 1982)

50 (Keown 1988, p. 408)

51 (Hafner and Hafner 1992) For information on these laboratory investigations, contact Mark S. Hafner at Louisiana State University email: namark@LSU.edu

52 (Gardner 1983, p. 5)

53 (Gardner 1999b, pp. 33–34)

54 (Gardner 1999b)

55 (Gardner 1999b, p. 170)

56 (Brewer 1997)

57 (Gendron 2000)

58 (Hazard 1998)

59 (Welch 1993)

60 (Brewer and Zabinski 1999)

61 (Offner 1994a; Offner 1994b)

62 (Gipps 1991)

63 (Riss 1993)

64 (Dolph and Dolph 1990)

65 (Matthews, C. 1996)

66 (Platt 1999)

67 (Shaw, Crocker, and Reed 1990)

68 (Guerrierie 1999)

69 (McCarty and Marek 1997)

70 (Lauer 2000)

71 (Peczkis 1993)

72 (Knapp and Thompson 1994)

73 (Chandler 1997)

74 (Goff 1995)

75 (Oden 1998)

76 (Scotchmoor 1995)

77 (Morishita 1991; Arbetman and Row 1985)

78 (Duveen and Solomon 1994; Solomon 1993)

79 (Gauld 1992)

80 (Lawson 1999)

81 (Seaford 1990)

References

Ackerman, P. D., & Williams, B. (1999). *Kansas Tornado: The 1999 Science Curriculum Standards Battle.* El Cajon, CA: Institute for Creation Research.

Alters, B. J. (1996). Letter to the editor. *The Chronicle of Higher Education, XLIII*(17), B3, B12.

Alters, B. J. (1999a). Reading stealth anti-evolutionary delivery systems: Possible effects on student science learning. *Reports of the National Center for Science Education, 19*(3), 27–29.

Alters, B. J. (1999b). What is creationism? *The American Biology Teacher, 61,* 103–106.

American Association for the Advancement of Science. (1993). *Benchmarks for Science Literacy.* New York: Oxford University Press.

Answers in Genesis. (no date [a]). *Creation Club Resources on the Web!* [Online]. Available: http://www.answersingenesis.org/home/area/creation_clubs/default.asp [2000, November 1].

Answers in Genesis. (no date [b]). *Creation Museum* [Online]. Available: http://www.answersin genesis.org/webman/article.asp?ID=3981 [1999, April 27].

Answers in Genesis. (2001). *Have you been blessed . . .* [Online]. Available: http://www.answersin genesis.org/ [2001, January 15].

Arbetman, L., & Row, R. L. (1985). *Great Trials in American History.* St. Paul, MN: West Publishing Co.

Association de Science Créationniste du Québec [Online]. Available: http://www.creation.on.ca/csic/creationnisme/accueil.htm [2000, November 25].

Audi, R. (1995). *The Cambridge Dictionary of Philosophy.* New York: Cambridge University Press.

Beem, K. (2000, August 2). Proevolution candidate wins education board primary. *The Kansas City Star* [Online] (n. pag.) Available: http//www.kcstar.com/item/pages/home.pat,local/3774a6cc.802,.html [2000, August 3].

Beeth, M. E. (1995). *Conceptual change instruction: Some theoretical and pedagogical issues.* Paper presented at the Annual Meeting of the National Association for Research in Science Teaching. (ED 407 267). Available from Educational Resources Information Center (ERIC) http://www.edrs.com/

Behe, M. J. (1996). *Darwin's Black Box: The Biochemical Challenge to Evolution.* New York: The Free Press.

Bell, G. (1997). *The Basics of Selection.* New York: Chapman & Hall/International Thomson Publishing.

Biola University. (no date). *Biola University Undergraduate On-line Application* [Online]. Available: http://www.biola.edu/admin/admissions/apply/eligibility.cfm [1998, August 16].

Bishop, B. A., & Anderson, C. W. (1986). *Evolution by natural selection: A teaching module.* (Occasional Paper No. 91). East Lansing: Institute for Research on Teaching, Michigan State University. Available from the Educational Resources Information Center (ERIC). ED 272 383.

Bishop, B. A., & Anderson, C. W. (1990). Student conceptions of natural selection and its role in evolution. *Journal of Research in Science Teaching, 27,* 415–427.

Brewer, C. A., & Zabinski, C. (1999). Simulating genetic change in a large lecture hall: The ultimate bean counting experience. *The American Biology Teacher, 61,* 298–302.

Brewer, S. D. (1997). *A model of desired performance in phylogenetic tree construction for teaching evolution.* Paper presented at the annual meeting of the National Association for Research in Science Teaching. (ED 405 216). Available from Educational Resources Information Center (ERIC) http://www.edrs.com/

Brumby, M. N. (1984). Misconceptions about the concept of natural selection by medical biology students. *Science Education, 68,* 493–503.

BSCS. (1992). *Evolution: Inquiries into biology and earth science* (annotated teacher's guide). Seattle: Videodiscovery.

Carroll, R. L. (1988). *Vertebrate Paleontology and Evolution.* New York: W. H. Freeman & Co.

Chandler, P. (1997). A missing link to understanding evolution. *Science Scope, 21*(2), 24–25.

Christensen, J. (1998, November 24). Teachers fight for Darwin's place in U.S. classrooms. *The New York Times* (Northeast edition), p. D3. Available: http://www.nytimes.com/library/national/science/112498sci-evolution.html [1998, November 26].

(The) Chronicle of Higher Education Career Network. (1998, November 27 issue). [Online]. Available: http://chronicle.com/weekly/jobs/faculty/sscience/education/93305.html [1998, November 21].

Clough, E. E., & Wood-Robinson, C. (1985). How secondary students interpret instances of biological adaptation. *Journal of Biological Education, 19,* 125–130.

Cole, J. (1995). NBC's mysterious extinction of facts. *Reports of the National Center for Science Education, 15*(4), 1, 9.

Darwin, C. (1898). *The Origin of Species by Means of Natural Selection.* Vol. II. New York: D. Appleton and Co.

Dawkins, R. (1987). *The Blind Watchmaker.* New York: W. W. Norton & Co., Inc.

Dawkins, R. (1996). *Climbing Mount Improbable.* New York: W. W. Norton & Co., Inc.

Deckard, S. W. (1998). A call to arms for conservative Christian science educators. *Impact* (ICR Series), No. 306, i–iv.

Demastes, S. S., Settlage, J., Jr., & Good, R. (1995). Students' conceptions of natural selection and its role in evolution: Cases of replication and comparison. *Journal of Research in Science Teaching, 32,* 535–550.

Desmond, A. J. (1997). *Huxley: From Devil's Disciple to Evolution's High Priest.* Reading, MA: Addison-Wesley.

Desmond, A. J., & Moore, J. (1991). *Darwin.* New York: W. W. Norton & Co.

Discovery Institute: Center for the Renewal of Science & Culture. (no date). *Intelligent Design Network, Inc. Announces a National Symposium!* [Online]. Available: http://www.discovery.org/comingEvents/darwinConference/index.html [2000, July 24].

Dobzhansky, T. (1967). Changing man: Modern evolutionary biology justifies an optimistic view of man's biological future. *Science, 155,* 409–415.

Dolph, G. E., & Dolph, L. L. (1990). A biological time capsule: Fossil fish. *The Science Teacher, 57*(8), 40–44.

Duschl, R. A., & Gitomer, D. H. (1991). Epistemological perspectives on conceptual change: Implications for educational practice. *Journal of Research in Science Teaching, 28,* 839–858.

Duveen, J., & Solomon, J. (1994). The great evolution trial: Use of role-play in the classroom. *Journal of Research in Science Teaching, 31,* 575–582.

Ehrenreich, B. (1999). The real truth about the female body. *Time* (Canadian Edition), *153*(9), 40–51.

Eldredge, N. (2000). *The Triumph of Evolution and the Failure of Creationism.* New York: W. H. Freeman & Company.

Eldredge, N., & Gould, S. J. (1972). Punctuated equilibria: An alternative of phyletic gradualism. In T. J. Schopf (Ed.), *Models in Paleobiology* (pp. 82–115). San Francisco: Freeman Cooper.

Focus on the Family. (1998a). Boundless Resource Center/Defeating Darwinism [Online]. Available: http://www.family.org/resources/itempg.cfm?itemid=772 [1999, July 3].

Focus on the Family. (1998b). Resource Center/Magazines/Newsletters [Online]. Available: http://www.family.org/resources/section.cfm?sid=47&pid=0 [1999, July 3].

Focus on the Family. (1998c). Teachers in Focus Magazine [Online]. Available: http://www.family.org/welcome/aboutfof/a0000083.html [1999, July 3]

Focus on the Family. (2000). About us/Meet our President [Online]. Available: http//www.family.org/welcome/ [2000, December 3].

Futuyma, D. J. (1995). *Science on Trial: The Case for Evolution*. Sunderland, MA: Sinauer Publishers.

Futuyma, D. J. (1998). *Evolutionary Biology* (3rd ed.). Sunderland, MA: Sinauer Publishers.

Futuyma, D. J. (Ed.) (2000). *Evolution, Science and Society: Evolutionary Biology and the National Research Agenda*. Piscataway, NJ: Office of University Publications, Rutgers, The State University of New Jersey.

Gallup, G., Jr., & Newport, F. (1991, November). Almost half of Americans believe biblical view of creation. *The Gallup Poll Monthly*, 30–34.

Gallup News Service. (2000, June 15). *Southern Baptists Pass New Statement of Faith Based on Literal Interpretation of Bible. Where Do Americans Stand?* [Online]. n. pag. Available: http://www.gallup.com/poll/releases/pr000615b.asp [2000, August 27].

Gallup Poll News Service. (1999, August 24–26). Gallup Poll Social Series: Labor and Education. (Data available from: The Gallup Organization, Inc., Princeton, NJ).

Gardiner, L. F. (1998). Why we must change: The research evidence. *Thought & Action: The NEA Higher Education Journal, 14*(1), 71–81.

Gardner, H. (1983). *Frames of Mind: The Theory of Multiple Intelligences*. New York: Basic Books.

Gardner, H. (1999a). *The Disciplined Mind: What all Students Should Understand*. New York: Simon and Schuster.

Gardner, H. (1999b). *Intelligence Reframed: Multiple Intelligences for the 21st Century*. New York: Basic Books.

Gauld, C. (1992). Wilberforce, Huxley & the use of history in teaching about evolution. *The American Biology Teacher, 54*, 406–410.

Gendron, R. P. (2000). The classification & evolution of caminalcules. *The American Biology Teacher, 62*, 570–576.

Gilchrist, G. W. (1997). The elusive scientific basis of intelligent design theory. *Reports of the National Center for Science Education, 17*(3), 14–15.

Gillespie, M. (1999, July 9). *Most Americans Support Prayer in Public Schools* [Online]. The Gallup Organization. n. pag. Available: http://www.gallup.com/poll/releases/pr990709.asp [1999, July 21].

Gingerich, O. (1992). Further reflections on "Darwin on Trial." *Perspectives on Science & Christian Faith, 44*, 253–254.

Gipps, J. (1991). Skulls and human evolution: The use of cases of anthropoid skulls in teaching concepts of human evolution. *Journal of Biological Education, 25*, 283–290.

Gish, D. T. (1974). *Have You Been . . . Brainwashed?* Seattle, WA: Life Messengers.

Gish, D. T. (1986). *Have You Been . . . Brainwashed?* (Rev. ed. 1994). Santee, CA: Gospel Tract Distributors.

Gish, D. T. (1995a). *Evolution: The Fossils STILL Say No!* El Cajon, CA: Institute for Creation Research.

Gish, D. T. (1995b). *Teaching Creation Science in Public Schools*. El Cajon, CA: Institute for Creation Research.

Godfrey, L. R., & Cole, J. R. (1986, August). Blunder in their footsteps: Creationists still cite the Texas "mantracks" as evidence that the dinosaurs and humans coexisted. *Natural History*, 4–12.

Goff, C. (1995). Survival of the fittest: Making a practical connection to Darwin's theory. *The Science Teacher, 62*(6), 24–25.

Gould, S. J. (1977). Evolution's erratic pace. *Natural History, 86*(5), 12–16.

Gould, S. J. (1978). Morton's ranking of races by cranial capacity. *Science, 200,* 503–509.

Gould, S. J. (1980). *The Panda's Thumb.* New York: W. W. Norton & Co.

Gould, S. J. (1989). *Wonderful Life: The Burgess Shale and the Nature of History.* New York: W. W. Norton & Co.

Gould, S. J. (1999). *Rocks of Ages: Science and Religion in the Fullness of Life.* New York: The Ballantine Publishing Group.

Gould, S. J., & Lewontin, R. C. (1979). The spandrels of San Marco: A critique of the adaptationist program. *Proceedings of the Royal Society of London, B 205,* 581–598.

Greene, E. D., Jr. (1990). The logic of university students' misunderstanding of natural selection. *Journal of Research in Science Teaching, 27,* 875–885.

Guerrierie, F. W. (1999). Beak adaptations. *Science Scope, 22*(4), 19–21.

Hafner, J. C., & Hafner, M. S. (1992). Laboratory investigations and discussions: An alternative pedagogical strategy in evolutionary biology. In R. G. Good et al. (Eds.), *Proceedings of the Evolution Education Research Conference* (pp. 115–123). Baton Rouge, LA: Louisiana State University.

Ham, K. (1987). *The Lie Evolution.* Green Forest, AR: Master Books, Inc.

Ham, K. (1998). *Creation Evangelism for the New Millennium.* Green Forest, AR: Master Books, Inc.

Ham, K. (2001, January). In the top 1% of the Web! *Answers Update, 3*(1), 7–8.

Harrison, A. G., Grayson, D. J., & Treagust, D. F. (1999). Investigating a grade 11 student's evolving conceptions of heat and temperature. *Journal of Research in Science Teaching, 36,* 55–87.

Hartwig, M. (1998). Skewed science. *Teachers in Focus* [Online]. n. pag. Available: http://www.family.org/cforum/teachersmag/features/a0002492.html [2000, December 3].

Hazard, E. B. (1998). Teaching about "intermediate forms." *The American Biology Teacher, 60,* 359–361.

Hemenway, R. E. (1999, October 29). The evolution of a controversy in Kansas shows why scientists must defend the search for truth. *The Chronicle of Higher Education* Section: Opinion & Arts, p. B7. Available: http://www.chronicle.com/weekly/v46/i10/10b00701.htm [2000, December 3].

Hung, E. H.-C. (1997). *The Nature of Science: Problems and Perspectives.* Belmont, CA: Wadsworth Publishing Co.

Huxley, T. H. (1882). *Science and Culture.* London, UK: Macmillan and Co.

Institute for Creation Research. (no date [a]). *Admissions Procedures* [Online]. Available: http://www.icr.org/grad/admiss.htm [1998, August 15].

Institute for Creation Research. (no date [b]). *Facilities* [Online]. Available: http://www.icr.org/abouticr/facility.htm [2000, August 19].

Institute for Creation Research. (no date [c]). *ICR Tenets of Creationism* [Online]. Available: http://www.icr.org/abouticr/tenets.htm [1998, August 15].

Institute for Creation Research. (1981). New book on Paluxy footprints published. *Acts & Facts, 10*(1), 1.

Institute for Creation Research. (1999). [Online]. Available: http://www.icr.org [1999].

Institute for Creation Research. (2000a). [Online]. Available: http://www.icr.org [2000, December 3].

Institute for Creation Research. (2000b). ICR on the cutting "edge:" New equipping course offered! *Acts & Facts, 28*(11), 1–2.

Institute for Creation Research. (2000c). ICR scientist conducts workshop on debating. *Acts & Facts, 29*(7), 10.

Institute for Creation Research. (2000d). ICR to dedicate new building at 30th anniversary celebration. *Acts & Facts, 29*(8), 3.

Institute for Creation Research. (2000e). Thirty years of creation evangelism and education. *Acts & Facts, 29*(10) [Online Issue No. 2]. n. pag. Available: http://www.icr.org/newsletters/afoct00 .html [2000, November 25].

InterVarsity Christian Fellowship/USA. (1999). *Frequently Asked Questions About InterVarsity* [Online]. Available: http://www.gospelcom.net/iv/general/faq.html [1999, June 27].

InterVarsity Christian Fellowship/USA. (2000). *Frequently Asked Questions About InterVarsity* [Online]. Available: http://www.gospelcom.net/iv/general/faq.html [2000, December 3].

InterVarsity Press. (1999). *Spring 1999 Academic Catalog*. Downers Grove, IL: Author.

Jensen, M. S., & Finley, F. N. (1995). Teaching evolution using historical arguments in a conceptual change strategy. *Science Education, 79,* 147–166.

Jensen, M. S., & Finley, F. N. (1996). Changes in students' understanding of evolution resulting from different curricular and instructional strategies. *Journal of Research in Science Teaching, 33,* 879–900.

Jensen, M. S., & Finley, F. N. (1997). Teaching evolution using a historically rich curriculum & paired problem solving instructional strategy. *The American Biology Teacher, 59,* 208–212.

Jimenez Aleixandre, M. P. (1994). Teaching evolution and natural selection: A look at textbooks and instructors. *Journal of Research in Science Teaching, 31,* 519–535.

Johnson, P. E. (1993). *Darwin on Trial* (2nd ed.). Downers Grove, IL: InterVarsity Press.

Johnson, P. E. (1995). *Reason in the Balance: The Case Against Naturalism in Science, Law, and Education.* Downers Grove, IL: InterVarsity Press.

Johnson, P. E. (1997). *Defeating Darwinism by Opening Minds.* Downers Grove, IL: InterVarsity Press.

Kaufman, M. (1998). Darwinists take their agenda to school. *Focus on the Family Citizen Magazine* [Online], n. pag. Available: http://www.family.org/cforum/citizenmag/departments/ a0001778.html [1999, July 3].

Kennedy, D. (1998, August 7). Helping schools to teach evolution. *The Chronicle of Higher Education,* p. A48.

Keown, D. (1982). Scaling our cosmic prisons. *The Science Teacher, 49,* 52–54.

Keown, D. (1988). Teaching evolution: Improved approaches for unprepared students. *The American Biology Teacher, 50,* 407–410.

King, M. (2000, July 27). Raëlians to start taking down names for peace. *The Gazette* (Montreal), p. A4.

Kinney, T. (2000, April 11). *Creationism Comes to Life: Ministry Dedicates Museum to Biblical Creation Account* [Online], n. pag. Available: http://more.abcnews.go.com/sections/science/dailynews/ creationisn_museum000417.htm [2000, June 9].

Kitcher, P. (1982). *Abusing Science: The Case Against Creationism.* Cambridge, MA: The MIT Press.

Knapp, P. A., & Thompson, J. M. (1994). Lessons in biogeography: Simulating evolution using playing cards. *Journal of Geography, 93*(2), 96–100.

Larson, E. J., & Witham, L. (1997). Scientists are still keeping the faith. *Nature, 386,* 435–436.

Laudan, L. (1983). The demise of the demarcation problem. In M. Ruse (Ed.), (1988). *But is it Science?* (pp. 337–350). Buffalo, NY: Prometheus Books.

Lauer, T. E. (2000). Jelly Belly® jelly beans & evolutionary principles in the classroom: Appealing to the students' stomachs. *The American Biology Teacher, 62,* 42–45.

Lawson, A. E. (1994). Research on the acquisition of science knowledge: Epistemological foundation of cognition. In D. L. Gabel (Ed.) *Handbook on Research on Science Teaching and Learning* (pp. 131–176). New York: Macmillan.

Lawson, A. E. (1999). A scientific approach to teaching about evolution & special creation. *The American Biology Teacher, 61,* 266–274.

Lawson, A. E., Abraham, M. R., & Renner, J. W. (1989). *A theory of instruction: Using the learning cycle to teach science concepts and thinking skills.* Manhattan, KS: National Association for Research in Science Teaching. (See http://www.narst.org/narst/mono.htm)

Lawson, A. E., & Weser, J. (1990). The rejection of nonscientific beliefs about life: Effects of instruction and reasoning skills. *Journal of Research in Science Teaching, 27,* 589–606.

Lawson, A. E., & Worsnop, W. A. (1992). Learning about evolution and rejecting a belief in special creation: Effects of reflective reasoning skill, prior knowledge, prior belief and religious commitment. *Journal of Research in Science Teaching, 29,* 143–166.

Lewontin, R. (2000). *It Ain't Necessarily So: The Dream of the Human Genome and Other Illusions.* New York: New York Review Books.

MacArthur, J. (1997). *The MacArthur Study Bible.* Nashville, TN: Word Bibles.

Marks, P. (1995). *Someone's Making a Monkey Out of You!* Colorado Springs, CO: Master Books.

Matsumura, M. (Ed.). (1995). *Voices for Evolution* (Rev. ed.). Berkeley, CA: The National Center for Science Education, Inc.

Matthews, C. E. (1996). Fossil finds. *Science Scope, 19*(7), 13–16.

Matthews, M. R. (1996). Editorial. *Science & Education, 5,* 91–99.

Maudlin, M. (1998, April 27). Inside CT: The more things change. *Christianity Today,* 4.

Mayr, E. (1997). *This is Biology: The Science of the Living World.* Cambridge, MA: The Belknap Press of Harvard University Press.

McCarty, R. V., & Marek, E. A. (1997). Natural selection in a petri dish. *The Science Teacher, 64*(8), 36–39.

McCourt, F. (1998, December). When you think of God what do you see? *Life,* 60–73.

McDowell, J. (1993). *The Best of Josh McDowell: A Ready Defense.* Nashville, TN: Thomas Nelson Publishers.

McDowell, J., & Stewart, D. (1980). *Answers to Tough Questions Skeptics Ask About the Christian Faith.* San Bernadino, CA: Here's Life Publishers Inc.

McGuire, J. E. (1992). Scientific change: Perspectives and proposals. In M. H. Salmon, J. Earman, C. Glymour, J. G. Lennox, P. Machamer, J. E. McGuire, J. D. Norton, W. C. Salmon, & K. F. Schaffner. *Introduction to the Philosophy of Science* (pp. 132–178). Englewood Cliffs, NJ: Prentice-Hall, Inc.

McInerney, J. (1997). Evolution of the NABT statement on the teaching of evolution. *Reports of the National Center for Science Education, 10*(1), 30–31.

Miller, K. R. (1999). *Finding Darwin's God: A Scientist's Search for Common Ground Between God and Evolution.* New York: Cliff Street Books/HarperCollins.

Moore, D. W. (1999, August 30). *Americans Support Teaching Creationism as Well as Evolution in Public Schools* [Online] The Gallup Organization. n. pag. Available: http://www.gallup.com/poll/releases/pr990830.asp [2000, August 27].

Moore, D. W. (2000, March 29). *Two of Three Americans Feel Religion Can Answer Most of Today's Problems* [Online]. The Gallup Organization. n. pag. Available: http://www.gallup.com/poll/releases/pr000329.asp [2000, August 27].

Moore, J. R. (1994). *The Darwin Legend.* Grand Rapids, MI: Baker Books.

Moore, R. (1999). Creationism in the United States VIII. The lingering threat. *The American Biology Teacher, 61,* 330–340.

Moreland, J. P. (1989). *Christianity and the Nature of Science.* Grand Rapids, MI: Baker Book House.

Morishita, F. (1991). Teaching about controversial issues: Resolving conflict between creationism & evolution through law-related education. *The American Biology Teacher, 53,* 91–93.

Morris, H. M. (1976). *The Genesis Record*. San Diego, CA: Creation-Life Publishers.

Morris, H. M. (1989). *The Long War Against God: The History and Impact of the Creation/Evolution Conflict*. Grand Rapids, MI: Baker Book House.

Morris, H. M. (1993). *Biblical Creationism: What Each Book of the Bible Teaches about Creation and the Flood*. Grand Rapids MI: Baker Books.

Morris, H. M. (1995a). *The Defender's Study Bible*. Grand Rapids, MI: World Publishing.

Morris, H. M. (1995b). The ICR outreach ministries IV: ICR, for such a time is this. *Back to Genesis* [Online], *82a*, n. pag. Available: http://www.icr.org/pubs/btg-a/btg-082a.htm [2000, December 3].

Morris, H. M. (Speaker). (1995c). *Biblical Creationism* (Cassette Recording No. 2). El Cajon, CA: ICR Summer Institute on Scientific Creationism.

Morris, H. M. (1997a). Defending the Faith. *Back to Genesis* [Online], *No. 97a* , n. pag. Available: http://www.icr.org/pubs/btg-a/btg-097a.htm [1998, November 22].

Morris, H. M. (1997b). *That Their Words May Be Used Against Them*. Green Forest, AR: Master Books.

Morris, H. M. (1998a). Bigotry in science. *Back to Genesis* [Online], *No. 114a*, n. pag. Available: http://www.icr.org/pubs/btg-a/btg-114a.htm [1998, November 22].

Morris, H. M. (1998b). Neocreationism. *Impact* [Online], *No. 296*, n. pag. Available: http://www.icr.org/pubs/imp/imp-296.htm [1998, December 20].

Morris, H. M., & Morris, J. D. (1996). *The Modern Creation Trilogy: Scripture and Creation*. Green Forest, AR: Master Books.

Morris, H. M., & Parker, G. E. (1982). *What is Creation Science?* San Diego, CA: Creation-Life Publishers.

Morris, J. D. (1976). The Paluxy River tracks. *Impact* (ICR Series), *No. 35*, i–viii. Available: http://www.icr.org/pubs/imp/imp-035.htm [2000, August 19].

Morris, J. D. (1980). *Tracking Those Incredible Dinosaurs: And the People Who Knew Them*. San Diego, CA: CLP Publishers.

Morris, J. D. (1990). Evolution and the wages of sin. *Impact* (ICR Series) [Online], *No. 209*, n. pag. Available: http://www.icr.org/pubs/imp/imp-209.htm [1998, November 20].

Morris, J. D. (1996). For such a time as this—ICR and the future—Part IX: Passing the mantle. *Back to Genesis, No. 87*, a–c.

Morris, J. D. (1999). *How Firm a Foundation in Scripture & Song*. Green Forest, AR: Master Books.

Morris, J. D. (1999, February). *Evangelism at ICR* [Online], n. pag. Available: http://www.icr.org/pubs/president/prz-9902.htm [1999, March 18].

Morris, J. D. (1999, March). Why do seminary professors entertain old earth ideas? *Back to Genesis No. 123b*, n. pag.

Morris, J. D., & Phillips, D. (1998). *Weapons of Our Warfare*. El Cajon, CA: Institute for Creation Research.

National Academy of Sciences. (1998). *Teaching About Evolution and the Nature of Science*. Washington, DC: National Academy Press.

National Academy of Sciences. (1999). *Science and Creationism: A View from the National Academy of Sciences* (2nd ed.). Washington, DC: National Academy Press.

National Research Council. (1996). *National Science Education Standards*. Washington, DC: National Academy Press.

National Science Board. (2000). *Science & Engineering Indicators—2000* (NSB-00-1). Washington, DC: U.S. Government Printing Office.

National Science Teachers Association. (1997). An NSTA position statement on the teaching of evolution. *Science Scope, 21*(2), 26–27.

Now Is The Time (Advertisement). (1997, April 28). *Christianity Today*. 27.

Numbers, R. L. (1992). *The Creationists*. New York: Alfred A. Knopf.

Numbers, R. L. (1998). *Darwinism Comes to America*. Cambridge MA: Harvard University Press.

O'Brien, T. (2000). A toilet paper timeline of evolution: 5 E cycle on the concept of scale. *The American Biology Teacher, 62*, 578–582.

Oden, D. (1998). Constructing a prehistoric adventure. *The Science Teacher, 65*(4), 38–41.

Offner, S. (1994a). Using chromosomes to teach evolution: I. Conserved genes & gene families. *The American Biology Teacher, 56*, 79–85.

Offner, S. (1994b). Using chromosomes to teach evolution: II. Chromosomal rearrangements in speciation events. *The American Biology Teacher, 56*, 86–93.

Osborne, R. J., & Wittrock, M. C. (1983). Learning science: A generative process. *Science Education, 67*, 489–508.

Overman, R., & Deckard, S. (1997). Origins beliefs among American science teachers (Revised). *Impact* (ICR Series) [Online], *No. 292*. Available: http://www.icr.org/pubs/imp/imp-292 .htm [1998, December, 12].

Overton. W. R. (1982). United States District Court Opinion: *McLean v. Arkansas*. In Ruse, M. (1988). *But Is It Science?* (pp. 307–331). Buffalo, NY: Prometheus Books.

Patterson, J. W. (1983). Thermodynamics and evolution. In L. R. Godfrey (Ed.), *Scientists Confront Creationism* (Chapter 6). New York: Norton.

Peczkis, J. (1993). Evolving student thought: Simulating evolution over many generations. *The Science Teacher, 60*(1), 42–45.

Pelikan, J. (Ed.) (1958). *Luther's Works: Volume 1. Lectures on Genesis, Chapters 1–5*. St. Louis, MO: Concordia Publishing House.

Pennisi, E. (2000, June 26). *Human Genome Spelled Out*. Available: http://sciencenow.sciencemag. org/cgi/content/full/2000/626/1. (This Web site is maintained by the American Association for the Advancement of Science [AAAS] and requires membership in the organization or a fee for use of the site.)

Pennock, R. T. (1999). *Tower of Babel: The Evidence against the New Creationism*. Cambridge MA: The MIT Press.

People for the American Way Foundation. (2000, March). *Evolution and Creationism in Public Education: An In-depth Reading of Public Opinion* [Online]. Available: http://www.pfaw.org/ issues/education/creationism-poll.pdf [2000, December 3].

Pinker, S. (1997a). The Big Bang. In M. Ridley (Ed.), *Evolution* (Chapter 58, pp. 353–366). Oxford, UK: Oxford University Press. (From *The Language Instinct*, Ch. II, 1994, New York: Morrow).

Pinker, S. (1997b). *How the Mind Works*. New York: W. W. Norton & Co.

Platt, J. E. (1999). Putting together fossil collections for "hands-on" evolution laboratories. *The American Biology Teacher, 61*, 275–281.

Posner, G. J., Strike, K. A., Hewson, P. W., & Gertzog, W. A. (1982). Accommodation of a scientific conception: Toward a theory of conceptual change. *Science Education, 66*, 211–227.

Raup, D. M. (1991). *Extinction: Bad Genes or Bad Luck?* New York: W. W. Norton & Co.

Rea-Ramirez, M. A., & Clement, J. (1998). *In search of dissonance: The evolution of dissonance in conceptual change theory*. Paper presented at the Annual Meeting of the National Association for Research in Science Teaching. (ED 417 985). Available from Educational Resources Information Center (ERIC) http://www.edrs.com.

Reasons to Believe. (no date). *Launch* [Online]. Available: http://www.reasons.org [1999].

Rechler, S. A. (1999, November 9). Scholars debate creationist theories. *Harvard Crimson,* 1, 3. Also available: *The Crimson Online* [Online], n. pag. http://www.thecrimson.com/news/printerfriendly.asp?ref=4450 [1999, November 22].

Riss, P. H. (1993). A ratio explanation for evolution. *Science Scope, 16*(4), 36–44.

Ross, H. N. (1994). *Creation and Time.* Colorado Springs, CO: NavPress Publishing Group.

Roth, K. J., Anderson, C. W., & Smith, E. L. (1987). Curriculum materials, teacher talk and student learning: Case studies in fifth grade science teaching, *Journal of Curriculum Studies, 19,* 527–548.

Ryrie, C. (1994). *The Ryrie Study Bible.* Chicago, IL: Moody Press.

Sadler, P. M. (1998). Psychometric models of student conceptions in science: Reconciling qualitative studies and distractor-driven assessment instruments. *Journal of Research in Science Teaching, 35,* 265–296.

Scharmann, L. C. (1993). Teaching evolution: Designing successful instruction. *The American Biology Teacher, 55,* 481–486.

Scotchmoor, J. (1995). The virtual museum of paleontology. *Science Scope, 18*(6), 50–54.

Scott, E. C. (1995). State of Alabama distorts science, evolution. *National Center for Science Education Reports, 15*(4), 10–11.

Scott, E. C. (1997). Antievolution and creationism in the United States. *Annual Review of Anthropology, 26,* 263–289.

Scott, E. C. (1998). [Review of the book *The Trouble with Science*]. *Reports of the National Center for Science Education, 18*(6), 25.

Scott, P. H. (1992). Conceptual pathways in learning science: A case study of one student's ideas relating to the structure of matter. In R. Duit, F. Goldberg, & H. Niedderer (Eds.), *Research in Physics Learning: Theoretical Issues and Empirical Studies* (pp. 203–224). Proceedings of an international workshop. Kiel, Germany: Institute for Science Education.

Seaford, H. W., Jr. (1990). Addressing the creationist challenge. *Anthropology & Education Quarterly, 21,* 160–166.

Settlage, J., Jr. (1994). Conceptions of natural selection: A snapshot of the sense-making process. *Journal of Research in Science Teaching, 31,* 449–457.

Shaw, E. L., Crocker, B., & Reed, B. (1990). Chipping away at fossils. *Science Scope, 14*(2), 30–31.

Shelley, M. (Ed.). (1994). *The Quest Study Bible.* Grand Rapids, MI: Zondervan Publishing House.

Shoshani, J. (1997). What can make a four-ton mammal a most sensitive beast? *Natural History, 106*(10), 37–44.

Simpson, G. G. (1952). How many species? *Evolution, 6*(1), 342.

Skehan, J. W., & Nelson, C. E. (2000). *The Creation Controversy & The Science Classroom.* Arlington, VA: NSTA Press.

Solomon, J. (1993). *Teaching Science Technology and Society.* Buckingham, England: Open University Press. ISBN 0–335–09952–1. Information: http://www.openup.co.uk/

Strickberger, M. W. (2000). *Evolution* (3rd ed.). Sudbury, MA: Jones & Bartlett Publishers.

Tobin, K., Tippins, D. J., & Gallard, A. J. (1994). Research on instructional strategies for teaching science. In D. L. Gabel. (Ed.) *Handbook of Research on Science Teaching and Learning* (pp. 45–93). New York: Macmillan Publishing Company.

Trowbridge, J. E., & Wandersee, J. H. (1994). Identifying critical junctures in learning in a college course on evolution. *Journal of Research in Science Teaching, 31,* 459–473.

Trowbridge, L. W., Bybee, R. W., & Carlson Powell, J. (2000). *Teaching Secondary School Science: Strategies for Developing Scientific Literacy.* Columbus, OH: Merrill.

U. S. Supreme Court Education Cases 2nd ed. (1991). Rosemount, MN: Data Research, Inc.

Van Campen, T. (1999, April 25). Creationism all over again. *The Gazette* (Montreal), p. C4.

Van Till, H. J. (1996). Basil, Augustine, and the doctrine of creation's functional integrity. *Science and Christian Belief, 8*(1), 21–38.

Vincenzo Bizzo, N. M. (1994) From down house landlord to Brazilian high school students: What has happened to evolutionary knowledge on the way? *Journal of Research in Science Teaching, 31,* 537–556.

Wandersee, J. H. (1985). Can the history of science help science educators anticipate students' misconceptions? *Journal of Research in Science Teaching, 23,* 581–597.

Wandersee, J. H. (1990). Concept mapping and the cartography of cognition. *Journal of Research in Science Teaching, 27,* 923–936.

Wandersee, J. H., Mintzes, J. J., & Novak, J.D. (1994). Research on alternative conceptions in science. In D. L. Gabel (Ed.), *Handbook on Research in Science Teaching and Learning* (pp. 177–210). New York: Macmillan.

Wehmeyer, P. (2000, June 9). *Net Religion: Millions are Exploring Spirituality Via the Internet* [Online]. Available: http://abcnews.go.com/onair/closerlook/wnt000609_cl_internetreligion_feature .html [2000, June 9].

Welch, L. A. (1993). A model of microevolution in action. *The American Biology Teacher, 55,* 362–365.

Wells, J. (2000). *Icons of Evolution: Science or Myth? Why Much of What We Teach About Evolution Is Wrong.* Washington, DC: Regnery Publishing, Inc.

Wheeler, D. L. (1996, November 1). A biochemist urges Darwinists to acknowledge the role played by an "Intelligent Designer." *The Chronicle of Higher Education,* p. A13.

Whitcomb, J. C. Jr., & Morris, H. M. (1961). *The Genesis Flood.* Grand Rapids, MI: Baker Book House.

Wilson, E. O. (1992). *The Diversity of Life.* Cambridge, MA: The Belknap Press of Harvard University Press.

Wilson, E. O. (2000). Vanishing before our eyes. *Time* (Special Edition Earth Day 2000) [Canadian Edition]), *155*(17), 29–31, 34.

Wilson, J. (1997, April 28). CT 97 book awards. *Christianity Today,* 12–13.

World Health Organization. (2000, June). *World Health Organization Report on Infectious Diseases 2000: Overcoming Antimicrobial Resistance.* Geneva, Switzerland: Author. (Publication code: WHO/CDS/2000.2).

Zuzovsky, R. (1994). Conceptualizing a teaching experience on the development of the idea of evolution: An epistemological approach to the education of science instructors. *Journal of Research in Science Teaching, 31,* 557–574.

Index